44 Structure and Bonding

Metal Complexes

With Contributions by
R. Bau, P. Gütlich, and R. G. Teller

With 104 Figures and 15 Tables

Springer-Verlag Berlin Heidelberg GmbH 1981

ISBN 978-3-662-15784-8   ISBN 978-3-540-38499-1 (eBook)
DOI 10.1007/978-3-540-38499-1

Library of Congress Catalog Card Number 67-11280

Table of Contents

Crystallographic Studies of Transition Metal Hydride Complexes

Raymond G. Teller* and Robert Bau*

Department of Chemistry, University of Southern California, Los Angeles, California 90007, U.S.A.

---

* Present address: Chemistry Division, Argonne National Laboratory, Argonne, Illinois 60439, U.S.A.

** Author to whom correspondence should be sent

A. Introduction

The importance of transition metal hydrides in modern inorganic chemistry cannot be overemphasized. Mononuclear, binuclear and polynuclear hydride complexes are plentiful, and many homogeneous catalytic systems are believed to involve some sort of transition metal hydride species as a reaction intermediate. Despite this, the ubiquitous hydrido complex is still probably the most poorly characterized of common organometallic complexes as far as accurate bond lengths and angles are concerned. The reason for this is easily appreciated if one recalls that X-rays are diffracted by electron density, and hence the scattering from a hydrogen atom is much weaker than the scattering from the larger atom to which it is bonded. However, accurate structural parameters are available from neutron diffraction studies, and it is to this technique that many investigators are gradually turning.

The first four decades of research on the chemistry and structures of transition metal hydride complexes have been thoroughly discussed in earlier reviews[1-3], and will not be repeated here. A comprehensive review of crystallographic work on hydride complexes up to 1970 has been published by Frenz and Ibers[1], and we ourselves have summarized various aspects of our neutron diffraction research in a few recent articles[4]. Additionally, a number of selective reviews[5,6] have appeared in a recent book[7] which summarizes the on-going research efforts of various groups active in metal hydride chemistry. In the present review it is our intent to list all transition metal hydride structures determined up to (and including) 1979. We have tried to be as thorough and complete as possible, but in the absence of a perfect searching and cataloguing system it is inevitable that some publications will have been inadvertently missed.

The earlier review by Frenz and Ibers[1] individually described all fifty-two structures known at the time. The number of structures reported since then (over three hundred) makes such an approach obviously impractical here. Instead, summaries of the various structural types will be presented, with a few representative examples selected for a brief discussion. The reader should realize that, because of the large number of molecules that need to be covered in this review, it will be virtually impossible to present an in-depth analysis of any individual compound.

In this article, the term "metal hydrides" refers almost exclusively to molecular species: that is, to covalent transition metal hydride complexes. We will not review the rather extensive literature on solid-state metal hydrides (e.g., UH_3, Th_4H_{15}, CuH, etc.)[8], nor that on metal hydride complexes of the non-transition metals (i.e., compounds with Al-H, Sn-H bonds, etc.).

I. Indirect Location of H Atoms Using X-Ray Data

In the introductory section of this article we will, out of convenience, take the liberty of using a few examples from our own experience to illustrate some of the methods that can be used to locate H atoms with x-ray data.

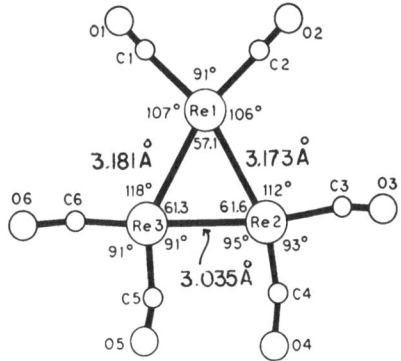

Fig. 1. A plot of the equatorial plane of the $[H_2Re_3(CO)_{12}]^-$ anion (Ref. 11). The long-long-short pattern of the Re—Re distances, coupled with the fact that the carbonyl groups on Re_2 and Re_3 seem to be "bending away" from Re_1, suggest the presence of bridging H atoms on the Re_1-Re_2 and Re_1-Re_3 edges. This represents an example of the use of molecular distortions to infer H positions in an X-ray study

The earliest attempts at determining the structures of transition metal complexes involved an indirect approach, in which H atom positions are inferred from the geometry of the remainder of the molecule. This technique is still heavily used today. Clues are usually derived from unusual distances and angles, or from the particular orientation of the other ligands in the complex. In the structure determinations of $HPtBr(PEt_3)_2$ [9] and $HMn(CO)_5$ [10], for example, a gap (or "hole") in the coordination sphere around the metal atom was interpreted to be a result of the steric influence of the "missing" H atom.

Often, angular distortions are combined with bond lengthening effects to deduce hydrogen positions. Take, for example, the structure of $[H_2Re_3(CO)_{12}]^-$ [11]. A plot of the equatorial plane (Fig. 1) shows that two of the Re-Re edges are longer than the third. Moreover, the C-Re-Re angles are distorted from their normal $105°$ values in a revealing fashion: the equatorial $M(CO)_2$ groups on $Re(2)$ and $Re(3)$ seem to be "bending away" from $Re(1)$. Both of these observations suggest that the H atoms are situated near the $Re(1)$-$Re(2)$ and $Re(1)$-$Re(3)$ edges of the triangle.

Sometimes the orientation of carbonyl groups can also give a clue about the hydrogen positions. In the case of $H_4Re_4(CO)_{12}$ [12], the carbonyl groups are eclipsed with respect to the M-M edges of the tetrahedron (*I*), in contrast to the usual staggered condition *II* [as found in $Ir_4(CO)_{12}$ [13] and $[H_6Re_4(CO)_{12}]^{2-}$] [14]. The disposition of the carbonyl groups in (*I*) is strongly indicative of the presence of face-bridging H atoms. Since $[H_6Re_4(CO)_{12}]^{2-}$ is believed to have edge-bridging hydrides [14], it is apparent that the $60°$ rotational difference of the $Re(CO)_3$ moiety in the two compounds is necessitated by the approximately octahedral $H_3Re(CO)_3$ coordination about rhenium.

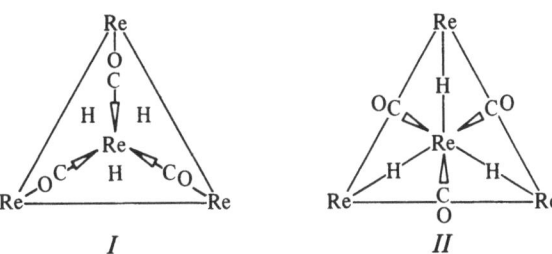

I   *II*

The use of differences in bond lengths is the most commonly-used method in the many situations in which H atoms are not unambiguously located in electron density maps. It is now fairly well established that single M—H—M bonds, in the absence of other bridging ligands, are longer than corresponding unbridged M—M bonds by about 0.10—0.45 Å, depending on the M—H—M angle. However, if another bridging group (such as PR_2, SR or CO) is present, the situation becomes ambiguous: in such cases the correlation of hydrogen bridging with metal-metal distance is no longer straightforward [15]. A coherent summary of the effect of bridging H atoms on M—M distances has been published by Churchill and co-workers [16].

II. Direct Location of H Atoms Using X-Ray Data

It is becoming increasingly common these days to find H atoms directly using X-ray diffraction methods. Success is not always guaranteed, but one's chances are generally better for hydride complexes of first row transition metals, as opposed to those of the heavier transition elements.

The first example of the direct location of H atoms is represented by the classic paper by La Placa and Ibers on $HRh(CO)(PPh_3)_3$ [17]. In this structure determination, the investigators overcame an unfavorable situation (trying to locate the lone hydrogen electron in the presence of the 45 rhodium electrons) by actually eliminating some data in the calculation of difference-Fourier maps. The reasoning was that since the scattering from hydrogen is largely confined to the low angle reflections (i.e., those occurring at small scattering angles), one should be able to enhance hydrogen peaks by removing the high-angle reflections from the calculation. By this technique, a very reasonable Rh-H distance of 1.60 (12) Å in $HRh(CO)(PPh_3)_3$ was determined.

The utility of La Placa's and Ibers' technique is illustrated in the structure determination of $H_3Mn_3(CO)_{12}$ [18] (Fig. 2). In this compound, the H atoms are suspected to lie in bridging positions in the equatorial plane. A conventional difference-Fourier map of this plane (Fig. 3a) shows a number of potential hydrogen peaks surrounded

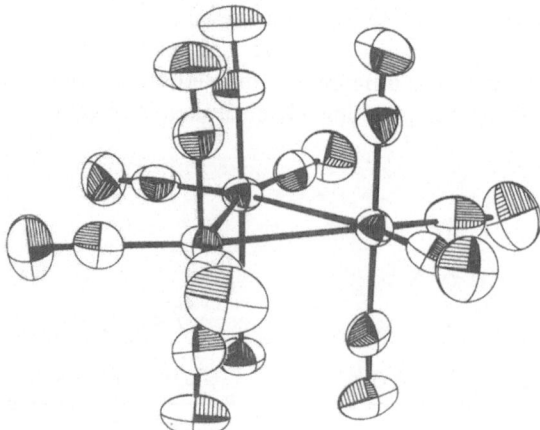

Fig. 2. The non-hydrogen framework of the $H_3Mn_3(CO)_{12}$ molecule (Ref. 18), as found prior to the location of the H atoms. The H atoms are believed to be situated in the equatorial plane in metal-metal bridging positions

by a large number of noise peaks. However, as successive shells of high-angle reflections are deleted from the calculation, the signal-to-noise ratio dramatically improves
and the map clears up considerably (Figs. 3b and c).

Although the La Placa/Ibers procedure has been criticized [19], subsequent experiments have shown that it is effective in many instances. One should not expect it to
work every time, but it does seem to improve significantly one's chances of finding
H atoms, and it is a useful technique for distinguishing "noise" from true H peaks. In
a recent publication, Dapporto and co-workers [20] have pointed out that Fourier
maps (as opposed to the conventionally-used *difference* Fourier maps) can sometimes
be more useful in revealing H positions, especially if the space group is centrosymmet-
ric.

Often, H coordinates determined from a difference-Fourier map are difficult to
refine in the subsequent least-squares process. In our experience, the success or
failure in the refinement can sometimes be critically dependent on the particular
choice of variables in the cycles immediately following the introduction of the H
positions [18].

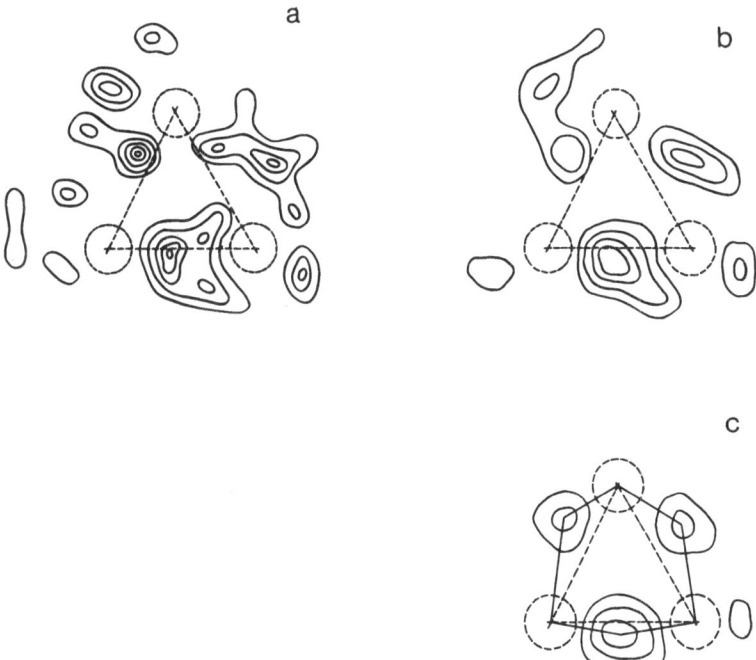

Fig. 3a–c. A series of difference-Fourier sections calculated through the equatorial plane of
$H_3Mn_3(CO)_{12}$ (Ref. 18). The dotted circles represent the positions of the Mn atoms which have
been subtracted out in this difference calculation. The sections correspond to calculations using
(a) all data (b) the inner two-thirds of the data and (c) the inner one-third of the data

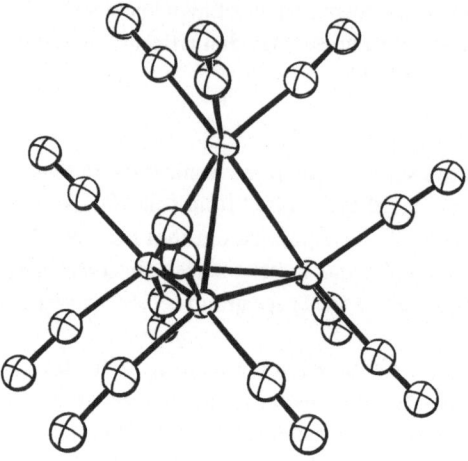

Fig. 4. The X-ray structure of H_4Re_4-$(CO)_{12}$ showing the non-hydrogen portion of the molecule (Ref. 12). Hydrogen atoms are believed to be situated in face-bridging positions

A few years ago we showed that in certain cases, a "Fourier-averaging" approach can yield good results. In the case of the face-bridging H atoms of $H_4Re_4(CO)_{12}$ [12] (Fig. 4), we were confronted with the problem of detecting one electron (from hydrogen) in the presence of 225 electrons (from three Re atoms), a much more unfavorable ratio than the Mn—H—Mn case discussed earlier (one electron in 50). Application of the high-angle-data-cutoff method yielded difference maps which were still ambiguous. It was reasoned, however, that if a molecule has several hydrogen positions and is of high symmetry, it may be possible to pool information from several different sections of the electron density map and come up with a usable composite. In such a composite section, random signals ("noise" peaks) would be expected to be cancelled out while true signals (hydrogen positions) would be enhanced. This technique can thus be thought of as an example of the signal-averaging concept as applied to X-ray crystallography. Accordingly, sections of the difference-Fourier map corresponding to the six mirror planes of the molecule were calculated and superimposed. Although the six individual sections were extremely "noisy" (Fig. 5), the composite section was remarkably "clean" (Fig. 6), with well-resolved peaks at the expected face-bridging positions and no electron density at the alternative edge-bridging or terminal positions. Admittedly, this approach has somewhat limited utility, since it is applicable primarily to molecules having high internal symmetry. Nevertheless, it illustrates that reasonable attempts can be made to "squeeze the last bit of information" from X-ray data.

Although the direct location of H atoms from X-ray data is often possible, success is notoriously unpredictable. In some cases H atom positions will be readily apparent from difference-Fourier maps, while in other cases one has to apply the "tricks" mentioned in this section to ferret them out. A good R index does not necessarily guarantee that the search for H atoms will be successful; conversely, structures with less impressive R indices will sometimes produce acceptable H positions [6]. Then there is the problem that H coordinates do not always converge during least-squares refinement. Even when they do, there is no guarantee that the refined positional parameters will be any better than the raw peak positions obtained from

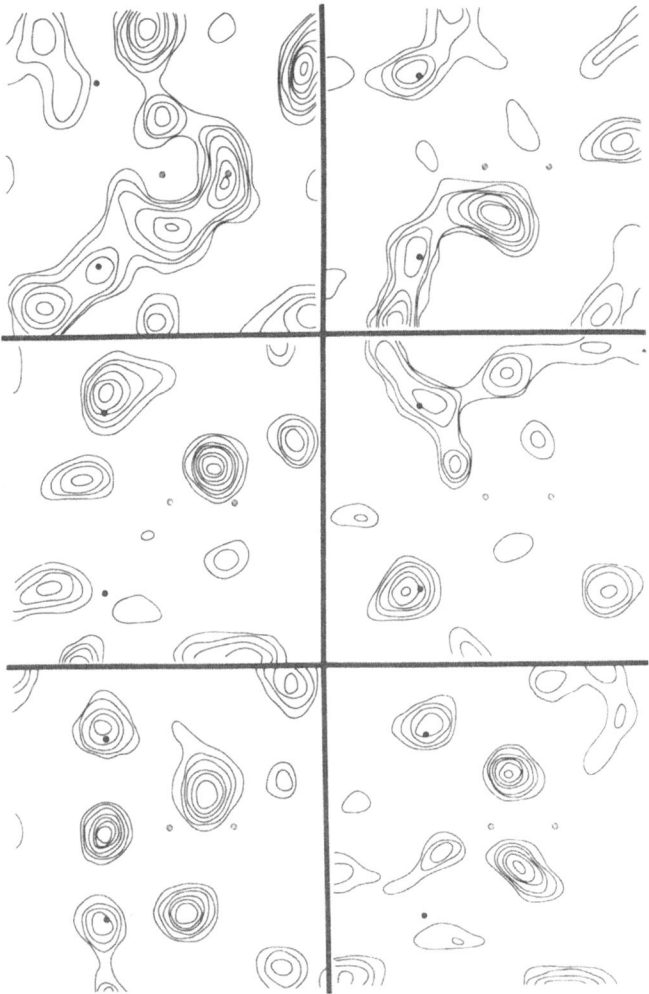

Fig. 5. Difference-Fourier maps corresponding to the six non-crystallographic mirror planes of the Re$_4$ tetrahedron. In each section, the two dots on the left represent Re atoms, the dot in the center represents the centroid of the tetrahedron, and the dot on the right corresponds to the mid-point of a Re—Re bond

the Fourier map. Despite these uncertainties, however, success does happen often enough to make the effort worthwhile.

Lastly, it should be pointed out that there are often significant differences between H positions determined by X-ray analysis and those determined from neutron data. Neutron diffraction provides true nuclear positions, whereas X-ray diffraction measures the electron density distribution. Thus, X-ray Fourier maps often give H peaks that, because of the perturbing influence of the M—H bonding electrons, appear closer to the M atoms than they really are [1]. A thorough analysis of this effect has

T

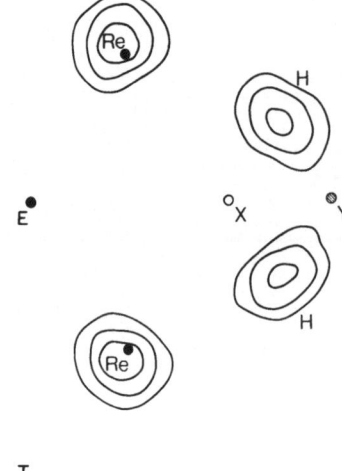

H

E

°X   ⊙Y

H

Re

T

Fig. 6. A composite difference-Fourier section for $H_4Re_4(CO)_{12}$, calculated by superimposing the maps in Fig. 5. Although the six individual sections contained many spurious peaks, in the composite shown here they are effectively averaged out. Point Y corresponds to the mid-point of a Re–Re bond (which is perpendicular to the plane of the paper), and point X represents the centroid of the tetrahedron. The H peaks shown here correspond to face-bridging positions, with a measured Re–H distance of 1.75–1.79 Å. The alternative edge-bridging position (marked E) and the terminal positions (marked T), which also lie on the mirror planes of the tetrahedron, show no electron density (Ref. 12)

been carried out by Petersen and Williams for the Mo–H–Mo bridge bond[21a]. In that paper, the X-ray[21a,c] and neutron[21b] results for $HMo_2Cp_2(CO)_4(PMe_2)$ were carefully compared. The H position, as determined from X-ray data, was found to be closer to the metal atom (by about 0.1 Å), making a more obtuse Mo–H–Mo angle.

III. Neutron Diffraction

Neutron diffraction[22] offers distinct advantages for investigators interested in determining light atom positions accurately in the presence of heavy atoms. Its main advantage arises from the fact that the neutron coherent scattering lengths (scattering factors) for most elements lie roughly within the same order of magnitude. In other words, neutrons are diffracted by most elements with more or less the same efficiency. To take a particular example, the X-ray scattering factors for the atoms in the molecule $HW_2(CO)_9(NO)$ are, roughly speaking, in the order of their atomic numbers: H,1; C,6; N,7; O,8; W,74. In contrast, the neutron scattering lengths for the same elements are (in units of 10^{-12} cm): H,–0.374; C,0.665; N,0.940; O,0.580; W,0.480. Note the following points:

(a) the magnitude of the neutron scattering length of tungsten is actually lower than those of carbon, nitrogen and oxygen;
(b) the sign of the scattering factor of H is negative, a point to which we will return later;
(c) the difference between C and N scattering is enhanced (from a 6:7 ratio for X-rays to a 2:3 ratio for neutrons), a fact which can be used to distinguish between C and N atoms in a molecule more easily than is possible with X-rays[23a].

But the main point is that the heavy atom clearly will not "swamp out" the H position under these circumstances, and, as a result, M—H distances can be determined much more accurately with neutrons than with X-rays (the difference in precision is typically one or two orders of magnitude).

Balanced against these positive features are two distinct disadvantages. Firstly, accessibility is limited: one obviously needs a source of neutrons, meaning that one has to carry out the experiment at a nuclear reactor or spallation source equipped for single-crystal diffraction work. At present, there are less than a dozen installations around the world equipped to carry out routine single-crystal neutron diffraction research. These include the Institut Laue-Langevin in Grenoble (France), the U.K. Atomic Energy Research Establishment in Harwell (England), and the Brookhaven (New York), Argonne (Illinois) and Oak Ridge (Tennessee) National Laboratories in the U.S.

The second limitation has to do with crystal size. In a single-crystal neutron diffraction experiment one usually needs crystals in the volume range $1-20 \, mm^3$, or about one thousand times the volume of crystals used in X-ray work. This is definitely a non-trivial problem for most transition metal hydride complexes. The principal reason for this requirement is that the monochromated particle flux from a typical neutron source is several orders of magnitude lower than the photon flux from a common X-ray tube. To compensate partially for this, a larger amount of sample is used. In addition, partly because of the low flux and partly because of the large bulk of most neutron diffractometers, neutron diffraction data are often collected at a rate that is 5 times or so slower than in X-ray diffraction. It is hoped that with the advent of high-intensity pulsed neutron spallation sources, such as the one currently being built at the Argonne National Laboratory[24], crystal size requirements and data collection times will be drastically reduced.

Another feature of neutron diffraction is the ability to distinguish between certain isotopes. The most striking difference is found between H and D, which have neutron scattering lengths ($\times 10^{-12}$ cm) of -0.374 and 0.667 respectively. The difference in sign causes H atoms to appear as negative peaks in neutron Fourier maps, and D atoms as positive peaks. In a recent study on the isotope distribution in a partially deuterated sample of $H_2Os_3(CO)_{10}(CH_2)$, it was found that the D atoms showed a preference for the carbon-bound (methylene) sites, while the H atoms preferred the osmium-bound (edge-bridging) sites[25]. Interestingly, the isotope ratio for one of the osmium-bound sites (64% H, 36% D) was such that the H and D contributions cancelled each other out completely. To our knowledge, this work represents the only example in which neutron diffraction was used primarily to distinguish between isotopes in an organometallic compound. Thus, the full potential of the method remains to be explored. One could imagine certain situations in which H/D discrimination would be useful: for example, in analyzing the labelled reactants and/or products of a stereospecific chemical reaction[26].

Certain elements pose problems for neutron diffraction work. Elements such as B, Cd, Sm, Eu and a few others are often difficult to work with because of their high absorption cross-sections for neutrons. Fortunately, with the exception of boron, these elements are not commonly encountered in organometallic complexes. For boron-containing compounds, there are two ways of getting around the absorption problem: either a sample is enriched in the low-absorbing isotope ^{11}B (as was done in the case of $B_{10}H_{14}$ [27]), or a complex having a low content of boron is selected: in the cases of $Hf(C_5H_4Me)_2(BH_4)_2$ [28] and $Cu(PPh_2Me)_3(BH_4)$ [29], for example, the mole fractions of boron are acceptably low (3.0% and 1.6% respectively).

There are a few other elements that are also poorly suited for neutron diffraction, but for a different reason: they scatter neutrons far too weakly. The best example is vanadium, whose neutron scattering length, -0.050×10^{-12} cm, is an order of magnitude smaller than those of most other elements. Thus, vanadium is to neutron diffraction what hydrogen is to X-ray diffraction: distances and angles involving it generally cannot be measured accurately in a neutron study. For a full listing of neutron scattering lengths and absorption factors of elements and their isotopes, the reader should consult Ref. 22.

Ideally, samples used for neutron study should be fully deuterated, because hydrogen has a large incoherent scattering amplitude which contributes considerable background scattering during data collection. Fully deuterated samples give data with superior peak-to-background ratios, which on principle should allow one to work with somewhat smaller crystals. However, in practice, investigators often do not bother to deuterate their samples, partly because of cost and also because, when data are obtained at low-temperature, the peak-to-background ratios are sufficiently high to allow very precise measurements of bond distances and angles.

Finally, we should mention a recent development that would allow investigators to get by with smaller-than-usual neutron data sets. Sheldrick and co-workers have developed a method of combined X-ray/neutron least-squares refinement[a] and have applied it successfully to several compounds, such as $H_2Os_3(CO)_{10}$ and $HOs_3(CO)_{10}$-(C_2H_3) [30]. The logic behind this approach is quite obvious: X-ray data contain a significant amount of information about heavy atom positions, neutron data about light atom positions; and hence it is advantageous to carry out the least-squares refinement with both sets of data simultaneously. The practical advantage of this idea is that it would allow investigators to derive maximum benefit from a neutron data set that is not large enough to solve the desired problem entirely on its own. Such a situation may arise if data collection time on the nuclear reactor is limited, or if the crystal is too small to give an adequate number of data.

IV. Other Methods for Estimating H Positions

There have appeared in the literature several reports on calculating probable H positions based on a consideration of intramolecular steric interactions between non-bonding atoms[b]. This method assumes that the geometry of the rest of the molecule (the non-hydrogen portion) is known.

---

[a]   A variation of this technique, in which data from two isotopically-substituted derivatives are combined, has long been used in powder neutron diffraction analysis. For example, a combined $HCrO_2/DCrO_2$ data set was used in the structure determination of chromous acid [31]

[b]   An analogous procedure, based on electrostatic potentials, was introduced earlier by Baur for H atoms in hydrogen bonds [32]

Albano and Bellon carried out potential energy computations for molecules of the type H_3ReL_4 (L = tertiary phosphine)[33], and found energy minima (for the H atoms) corresponding to equatorial positions in a distorted pentagonal bipyramidal arrangement. In this work, stereochemical conclusions were derived mainly from an analysis of H ... H contacts. This method was later applied to $[HNi_2(CO)_6]^-$ [34], $HRe(NO)(PPh_3)_2$ [35] and $[H_4Re_4(CO)_{13}]^{2-}$ [36].

More recently, Orpen has carried out potential energy calculations for a large number of metal carbonyl cluster hydrides[37]. Agreement between calculated and observed H positions, in several complexes with accurately known hydride configurations, turned out to be quite good (with all predicted positions coming within 0.1 Å of known H coordinates), and predictions were made concerning the still-unknown hydride positions in $H_2Os_6(CO)_{18}$ and $H_3Os_4(CO)_{11}(C_6H_9)$.

Another idea that has been tried is based on a fragment-fitting approach. In the X-ray structure determination of $[H_2W_2Cp_4(\mu\text{-}H)]^+$ [38], the known geometry of H_2MoCp_2 (determined previously from neutron diffraction[39]) was superimposed on the X-ray-visible WCp_2 portion of $[H_2W_2Cp_4(\mu\text{-}H)]^+$ to yield reasonable positions for the terminal and bridging hydrogen atoms of the dimer.

Sheldrick and Yesinowski[40] have recently applied nematic phase nmr data to elucidate the H positions in $H_3Ru_3(CO)_9(CCH_3)$, using the heavy atom positions derived from X-ray data. It is in principle an accurate method which does not require large crystals: the only difficulties are that many hydrides react with nematic liquid crystals, and that some molecular symmetry is desirable[41].

Finally, mention should be made of a recent article by Green, Mingos and Seddon[42], who have extended Lipscomb's "styx" rules[43] (developed for the boron hydrides) to metal hydride cluster complexes, and have shown that they correctly predict the different configurations of H atoms in $H_4Re_4(CO)_{12}$ (face-bridging) and $H_4Ru_4(CO)_{12}$ (edge-bridging).

B. Terminal M–H Bonds

The first metal hydride complex studied by neutron diffraction, and still perhaps the most intriguing one, is K_2ReH_9, described in the classic paper by Abrahams, Ginsberg and Knox[44]. The beautiful tricapped trigonal prismatic symmetry of the $[ReH_9]^{2-}$ anion and the evolution of the compound's chemical formula (from KRe to $KReH_4$, K_2ReH_8, and finally K_2ReH_9) have been discussed in earlier reviews[1-3].

Of the monomeric metal carbonyl hydrides $[HMn(CO)_5, H_2Fe(CO)_4, HCo(CO)_4]$, only $HMn(CO)_5$ has been analyzed crystallographically[10,45]. The lack of results for $H_2Fe(CO)_4$ and $HCo(CO)_4$ is probably related to the fact that the compounds are thermally unstable. The neutron diffraction analysis of $HMn(CO)_5$ by La Placa and co-workers[45] produced a Mn–H distance of 1.60(2) Å. This result, together with the earlier measurement of the Re–H distance of 1.68(1) Å in K_2ReH_9, provided conclusive proof that M–H distances in metal hydride complexes are "normal" (i.e., are consistent with the known covalent radii of the elements), and not anomalously short as suggested by some researchers (this controversy is reviewed in Refs. 1–3).

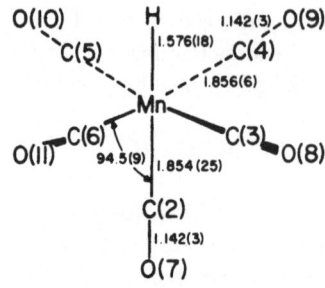

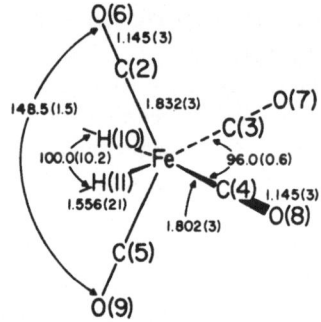

H
1.556(18)
O(7)   O(8)
        C(4)   172.6(1.2)   O(9)
    C(3)   Co   1.818(3)   C(5)
                    99.7(6)
1.764(10)

C(2)
1.141(3)
O(6)

Fig. 7. The structures of $HMn(CO)_5$, $H_2Fe(CO)_4$, and $HCo(CO)_4$ as determined by gas-phase electron diffraction (Ref. 46)

Although $H_2Fe(CO)_4$ and $HCo(CO)_4$ have not been studied crystallographically, they (together with $HMn(CO)_5$) have been analyzed with gas-phase electron diffraction techniques [46]. The geometries which fit the data best are illustrated in Fig. 7. The Mn–H bond length, 1.58(2)Å, closely matches the earlier-mentioned neutron diffraction results on the same complex [1.60(2)Å] [45]. Similarly, the Fe–H distance in $H_2Fe(CO)_4$ (average = 1.56(2)Å) agrees well with the results of an X-ray investigation of $H_2Fe[P(OEt)_2Ph]_4$ [Fe–H = 1.51(4)Å] [47]. Both of these iron complexes have the *cis* configuration and are appreciably distorted from the ideal octahedral geometry. The geometry of $HCo(CO)_4$ closely resembles that of the isoelectronic $[HFe(CO)_4]^-$, which had been analyzed earlier with X-ray diffraction [48]. They have a structure based on a distorted trigonal bipyramid, with the equatorial CO groups appreciably bent towards the axial H ligand: Co–H = 1.56(2)Å and C(ax)–Co–C(eq) = 99.7(6)° in $HCo(CO)_4$; Fe–H = 1.57(12)Å and C(ax)–Fe–C(eq) = 99.1(7)° in $[HFe(CO)_4]^-$.

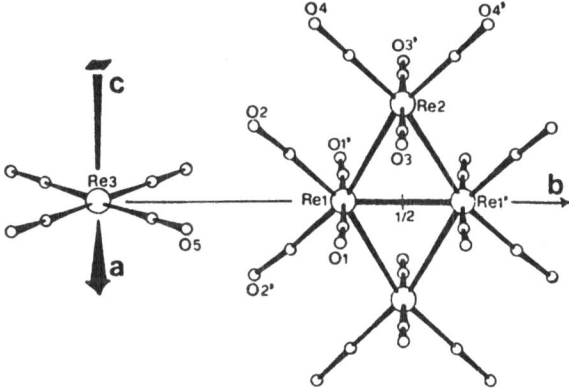

Fig. 8. The *trans* isomer of $[H_2Re(CO)_4]^-$, shown on the *left* without its H atoms (which were not located). This complex was isolated as an unusual mixed salt $[NEt_4]^+[H_2Re(CO)_4]^- \cdot 2[NEt_4]_2^+[Re_4(CO)_{16}]^{2-}$, and its companion $[Re_4(CO)_{16}]^{2-}$ anion is shown on the *right* (Ref. 50)

Other monomeric carbonyl hydrides whose structures have been investigated are the *cis* and *trans* isomers of $[H_2Re(CO)_4]^-$ [49,50]. In neither case have the hydride ligands been directly located. *Trans*-$[H_2Re(CO)_4]^-$ was prepared in low yield from the pyrolysis of *cis*-$[H_2Re(CO)_4]^-$, and was isolated as crystals of an unusual mixed salt $[NEt_4]_2^+[Re_4(CO)_{16}]^{2-} \cdot \frac{1}{2}[NEt_4]^+[H_2Re(CO)_4]^-$ [50] (Fig. 8). The *cis* and *trans* isomers of $[H_2Re(CO)_4]^-$, together with the *fac* and *mer* forms of H_3IrL_3 (L = tertiary phosphine) [51,52], provide rare examples of stereochemically rigid polyhydride complexes. Most other polyhydride complexes (especially those that are not six-coordinate) tend to be fluxional [53], and it is perhaps the special stability of the octahedral arrangement that enables $[H_2Re(CO)_4]^-$ and $H_3Ir(PR_3)_3$ to exist in isomeric forms.

Accurate neutron diffraction measurements are now available for the terminal M—H distances in H_2MoCp_2 [Mo—H = 1.685(3)Å] [39], H_3TaCp_2 [average Ta—H = 1.774(3)Å] [54], and the binuclear bent sandwich complex $H_4Th_2(C_5Me_5)_4$ [Th—H = 2.03(1)Å] [55]. H_2MoCp_2 is of particular interest because of the early controversy regarding the disputed location of its H atoms [56]. The geometry of H_3TaCp_2 (Fig. 9) is characterized by an exceptionally "tight" H_3Ta central fragment [H—Ta—H = 62.9(4)°, H ... H = 1.851(9)Å] [54]. The structure determination of $H_4Th_2(C_5Me_5)_4$ is noteworthy in that it represents the first time the structure of an organometallic molecule was solved directly from neutron data; i.e., without carrying out an X-ray analysis first. Its Th—H distance [2.03(1)Å] is the longest terminal M—H bond length known. Another noteworthy metallocene structure is illustrated in Fig. 10 for "niobocene" [57]. Unlike most other metallocenes, which are monomeric sandwich compounds, "niobecene" is dimeric $[H_2Nb_2(C_5H_5)_2(C_5H_4)_2]$. Each Nb atom has inserted itself into a C—H bond, with the result that the ligands around Nb are π-C_5H_5, π-C_5H_4, σ-C_5H_4, σ-H. A similar situation is seen in the tungsten analogue, $[H_2W_2(C_5H_5)_2(C_5H_4)_2]$ [58], which however has a *trans* configuration of the Cp rings as opposed to the *cis* arrangement found in the niobium dimer.

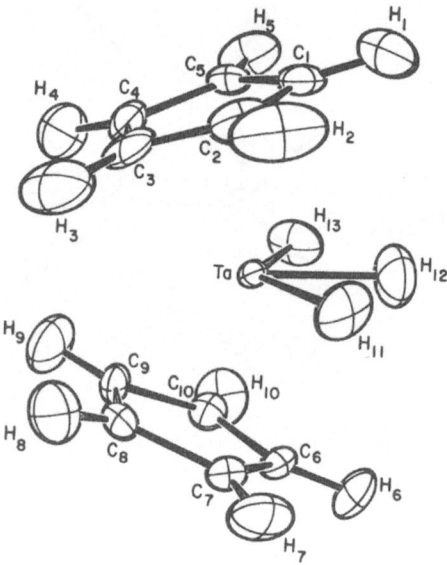

Fig. 9. The structure of H_3TaCp_2 (neutron study, Ref. 54). The equatorial H_3Ta fragment, which is planar within experimental error, serves as a non-crystallographic mirror plane for the molecule

The structures of a rather large number of mononuclear hydride/phosphine complexes (e.g., H_7ReL_2, H_5ReL_3, H_3ReL_4, H_4OsL_3, H_3IrL_3, etc., where L = tertiary phosphine) have recently been reviewed by us[51] and will not be repeated here. We will simply mention that one of these compounds, $H_4Os(PMe_2Ph)_3$, has been analyzed by neutron diffraction[59] and shows a distorted pentagonal bipyramidal geometry with the four hydride ligands [average Os–H = 1.659(3) Å] in the equatorial plane (Fig. 11). The structure of a particularly fascinating dimeric member

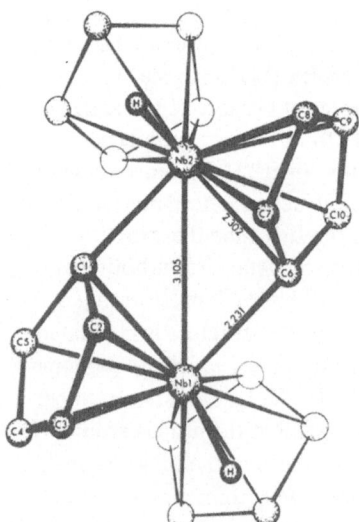

Fig. 10. The structure of a molecule commonly called "niobocene", whose actual formula is $[HNb(C_5H_5)(C_5H_4)]_2$ (Ref. 57). Unlike most other metallocenes, "niobocene" is dimeric

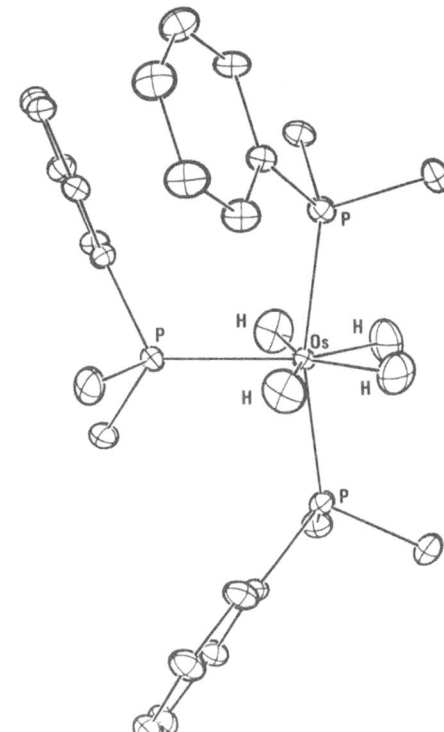

Fig. 11. The structure of $H_4Os(PMe_2Ph)_3$, as determined by neutron diffraction (Ref. 59). The geometry is based on a distorted pentagonal bipyramid, with an essentially planar H_4OsP equatorial moiety. Note the slight bending of the axial phosphine ligands away from the equatorial phosphine

of this family, $H_8Re_2(PEt_2Ph)_4$ [60)], will be presented later in this article in connection with M—H—M bridged complexes (Sect. D. III.).

In virtually all mononuclear hydride complexes, the H ligand occupies a stereochemically active position: in other words, it is usually possible to infer H positions by searching for "holes" in the coordination sphere around the metal atom. Exceptions, however, are found in the structures of $HCo(PF_3)_4$ [61)], $HRh(PPh_3)_4$ [62)], and $HRh(PPh_3)_3(AsPh_3)$ [63)]. In these complexes, the P and As atoms define almost tetrahedral coordination spheres around the central atom (Fig. 12). Under these circumstances, it is not obvious from X-ray results alone where the H atoms might be located.

A series of cobalt and nickel hydrido complexes with tetradentate ligands of the general formula $E(C_2H_4PPh_2)_3$ or $E(o\text{-}C_6H_4PPh_2)_3$ (where E = P or N) have been studied [64—69)]. Generally, these structures are trigonal bipyramidal with a lone hydrogen ligand occupying an axial position. In $HCoP(C_2H_4PPh_2)_3$ [65)], the hydrogen ligand was located and successfully refined [Co—H = 1.43(6) Å]. Interestingly, the coordination geometries of these complexes very closely approximate those of ideal trigonal bipyramids, unlike other hydrido complexes which almost invariably show distortions owing to the low steric requirements of the hydrogen atom. This lack of distortion is most probably due to the steric constraints within the tetradentate ligand. Another member of the series, $[H_xNiN(CH_2CH_2PPh_2)_3]^+[BF_4]^-$, is espe-

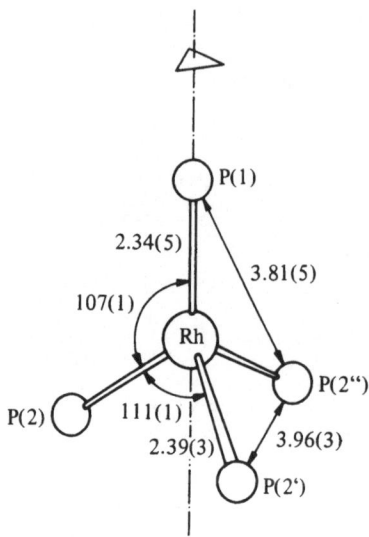

Fig. 12. The geometry of HRh(PPh$_3$)$_4$ (Ref. 62), which shows very little distortion from tetrahedral symmetry. This is an example of a molecule whose hydrogen location is not at all obvious

cially intriguing since crystals of it are believed to contain a non-stoichiometric amount of hydrogen [69]. Solid samples can be isolated having H:Ni ratios ranging from 0.04 to 0.83, and the particular crystal structure determination was carried out on a sample having H:Ni = 0.5. It is perhaps the only example of a molecular compound where non-stoichiometric hydrogen/metal ratios can be found in the solid state. This phenomenon, however, is commonly found in binary metal hydrides [8].

Several compounds are known having an H atom and an organic group attached to the same metal atom. Specific examples include HPt(CH$_2$CN)(PPh$_3$)$_2$ [70] and HReCp(CO)$_2$(CH$_2$Ph) [71] (both of which contain a hydride/alkyl grouping), HRh(PPh$_3$)[C$_2$B$_9$H$_{10}$-C$_2$H$_4$CH=CH$_2$] (hydride/alkene) [72], HIr(Cl)(η^3-C$_3$H$_4$Ph)-(PPh$_3$)$_2$ (hydride/allyl) [73], and [HRu(C$_4$H$_6$)(PMe$_2$Ph)$_3$]$^+$[PF$_6$]$^-$ (hydride/diene) [74]. The stability of HIr(Cl)(η^3-C$_3$H$_4$Ph)(PPh$_3$)$_2$ has been attributed to steric interactions between the phenyl rings of the allylic group and the triphenylphosphine ligands, which prevent the formation of the corresponding 16-electron olefinic complex [73].

A number of other structure determinations are worthy of mention. One is the interesting complex "[HRu(PPh$_3$)$_3$]$^+$[BF$_4$]$^-$" [75], which appears at first glance to be a 14-electron species. The X-ray structural analysis, however, established that one of the triphenylphosphine ligands is bound to the Ru atom via a phenyl ring (Fig. 13). Hence this molecule is better formulated as [HRu(PPh$_3$)$_2$(η^6-C$_6$H$_5$-PPh$_2$)]$^+$[BF$_4$]$^-$. Another apparently electron-deficient compound with the empirical formula Ru(Me$_2$PC$_2$H$_4$PMe$_2$)$_2$ was found to be a dimeric hydrido complex with an unusual Ru–P–C–Ru bridge [76]. Each metal atom has been inserted into a C–H bond of a neighboring Me$_2$PC$_2$H$_4$PMe$_2$ ligand, resulting in the formation of a dimer. Finally, the unusual ruthenium hydrido complex HRu(PPh$_3$)$_3$[HC=C(Me)C(O)OC$_4$H$_9$] is formed in the reaction between H$_2$Ru(PPh$_3$)$_4$ and n-butylmethacrylate [77]. The Ru atom has oxidatively added to a vinylic C–H bond, and this compound represents the first example of such a structure.

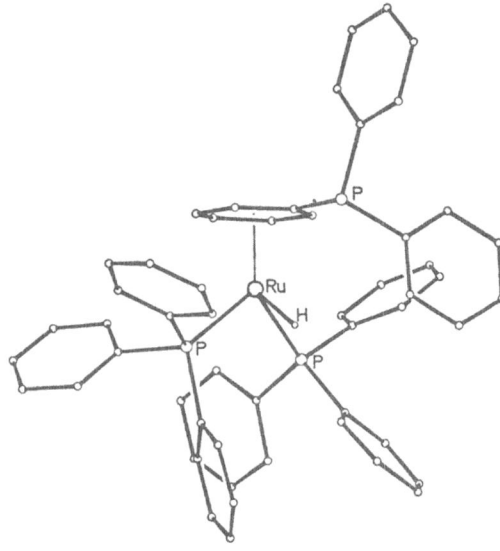

Fig. 13. In the $[HRu(PPh_3)_3]^+$ cation, two phosphines are bonded to the metal atom conventionally, while the third is π-bonded through a phenyl ring (Ref. 75)

Terminal M–H bonds in metal cluster complexes are occasionally found (see Table 3). Some of them are noteworthy in that they contain both terminal and bridging H ligands on the same metal atom. Examples include $H_2Os_3(CO)_{11}$ [78] (Fig. 14) and $[H_4Re_4(CO)_{15}]^{2-}$ [79] (which contain edge-bridging and terminal H's), $[H_7Ir_3(py)_3(PCy_3)_3]^{2+}$ [80] (which contain face-bridging and terminal H's), and $\{H_7Ir_3[Ph_2P(CH_2)_3PPh_2]_3\}^{2+}$ [81] (which contain all three types of H ligands).

A series of fascinating solid-state ternary complexes have been reported in the literature: compounds containing (i) an alkali or alkaline-earth metal, (ii) a transition metal, and (iii) hydride ligands. The preparation and properties of these compounds, of which a few dozen are known, have recently been reviewed by Moyer, Lindsay and Marks [82]. Representative examples are Li_4RhH_4, Sr_2RuH_6, and Mg_2NiH_4. Interest in these compounds, for the purposes of this review, lies in the fact that there are indications that covalent M–H interactions might exist in some of them. For example, powder neutron diffraction studies on Sr_2RuD_6 [83] revealed

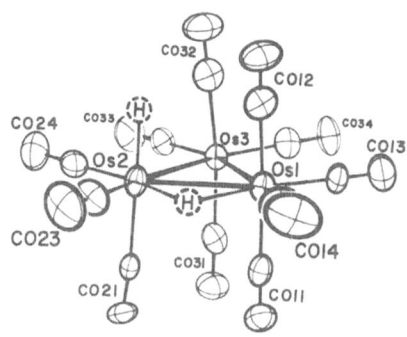

Fig. 14. The structure of $H_2Os_{11}(CO)_{11}$, a rather uncommon example of a cluster complex containing both terminal and bridging H ligands attached to the same metal atom (Ref. 78)

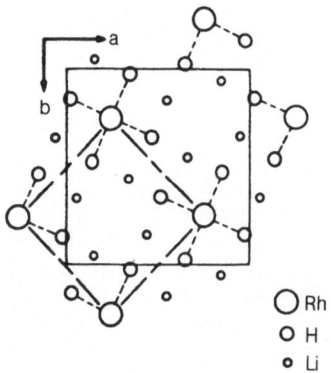

O Rh
O H
o Li

Fig. 15. A projection of the unit cell of Li_4RhH_4, showing the square-planar RhH_4 units (Ref. 84)

discrete octahedral RuD_6 units with a Ru—D distance of 1.69 Å, a distance attributable to a covalent interaction. Similarly, single-crystal X-ray work on Li_4RhH_4 [84] shows the presence of square planar RhH_4 units (Fig. 15). It is interesting to speculate if an entire series of polyhydridic anions might exist in these solids (e.g., $[RuH_6]^{4-}$, $[RhH_5]^{4-}$, $[NiH_4]^{4-}$, etc.), that would be isoelectronic with the well-known $[ReH_9]^{2-}$ anion.

C. M—H—X Bonds

I. M—H—B Systems

1. BH₄ Complexes

An excellent comprehensive review of the chemistry and structures of metal-BH_4 complexes has been published by Marks and Kolb [85]. The attachment of a BH_4 ligand to a metal atom can assume either unidentate (*III*), bidentate (*IV*) or tridentate (*V*) forms, which are often distinguishable from each other on the basis of M—B distances [86-89].

III   *IV*   *V*

In addition, for metal complexes of the larger boron hydrides, more complicated types of bonding (where M—H—B linkages are augmented by M—B, M—M or B—B bonds) are found. As mentioned earlier in this article, metalloborane complexes are generally unsuitable for neutron diffraction because of the very large absorption cross

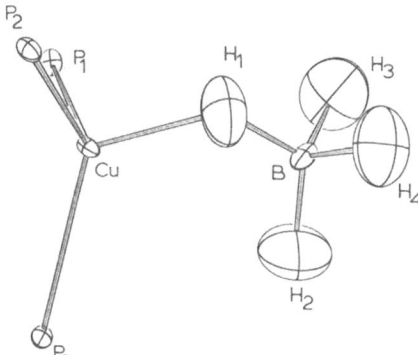

Fig. 16. The P_3CuBH_4 core of $Cu(BH_4)$-$(PPh_2Me)_3$, as determined by neutron diffraction (Ref. 29). This is the only example of a unidentate BH_4 complex structurally characterized. The Cu–H–B linkage is distinctly bent $[121.7(4)°]$

section of boron for neutrons. However, if the amount of boron in the sample is low (ideally, a mole fraction of less than 5%), absorption problems are not severe.

Of the three modes of metal/BH_4 bonding depicted above, the unidentate type is by far the least common. Only one compound of this class has been structurally characterized: $Cu(PPh_2Me)_3(BH_4)$ (Fig. 16), which has been studied by X-ray [90] and neutron [29] methods. In this molecule, the Cu–H–B bond is distinctly bent $[Cu–H–B = 121.7(4)°]$, a characteristic feature which has been found in all compounds having unsupported 3-center/2-electron bonds [4]. The distances in the M–H–B core of this molecule are as follows: Cu–H = 1.697(5) Å, B–H = 1.170(5) Å, Cu–B = 2.518(3) Å. Another interesting feature of the structure of $Cu(PPh_2Me)_3(BH_4)$ is the fact that a second H atom, labelled H(2) in Fig. 16, is situated 2.722(7) Å from

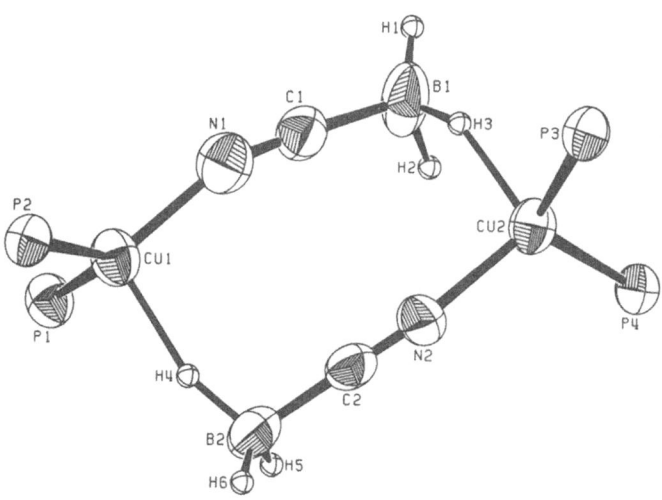

Fig. 17. The molecular geometry of $[(PPh_3)_2CuNCBH_3]_2$, with the phenyl rings omitted for clarity. (Ref. 91)

the Cu atom. This distance, while non-bonding, is close enough to suggest a potential Cu–H bond about to be formed (indeed, it is known that when the compound is dissolved, one of the phosphine ligands dissociates and the BH_4 ligand becomes bidentate). A somewhat related structure having unidentate B–H–M bridges has been reported for $[Cu(PPh_3)_2(NCBH_3)]_2$ [91] (Fig. 17). However, in this case the B–H–Cu bond is supplemented by N–Cu bonding from the cyano end of the ligand, generating a head-to-tail dimer. The average Cu–H–B angle was reported to be $153(7)°$ in $[Cu(PPh_3)_2(NCBH_3)]_2$.

Bidentate BH_4 ligands are very common. The X-ray analysis of several molecules having this structural feature have been carried out. Examples include $[Mo(CO)_4(BH_4)]^-$ [92], $HM(PCy_3)_2(BH_4)(M = Co, Ni)$ [93,94], $Co[MeC(CH_2PPh_2)_3](BH_4)$ [95], and $M(C_5H_5)_2(BH_4)(M = Ti, Nb)$ [96,97]. One of these papers [93a] contains a concise review of the structures of molecules containing $M(\mu\text{-}H)_2BH_2$ ligands. Very recently, neutron diffraction results on $Hf(C_5H_4Me)_2(BH_4)_2$ [28] (Fig. 18) and $Co(terpyridine)$ (BH_4) [98] became available. In the former case, the following distances and angles for the $Hf(\mu\text{-}H)_2B$ portion of the molecule were obtained: Hf–H = $2.069(7)$, $2.120(8)$ Å, B–H = $1.255(9)$, $1.208(13)$ Å, Hf–B = $2.553(6)$ Å, Hf–H–B = $96.3(5)$, $97.3(5)°$.

The tridentate mode of attachment is also rather common. The first structural report of this arrangement appeared in 1967, when the X-ray analysis of $Zr(BH_4)_4$ (Fig. 19) was described by Bird and Churchill [99]. However, the twelve bridging H atoms could not be located in that study. The correctness of the tridentate-bridged model was later confirmed by two independent gas-phase electron diffraction investigations [100], which came up with measurements of Zr–H = $2.21(4)$ Å, B–H = $1.27(5)$ Å, and Zr–B = $2.31(1)$ Å for the M–H–B bridging portion of the molecule. The analogous molecule $Hf(BH_4)_4$ has been analyzed twice by neutron diffraction:

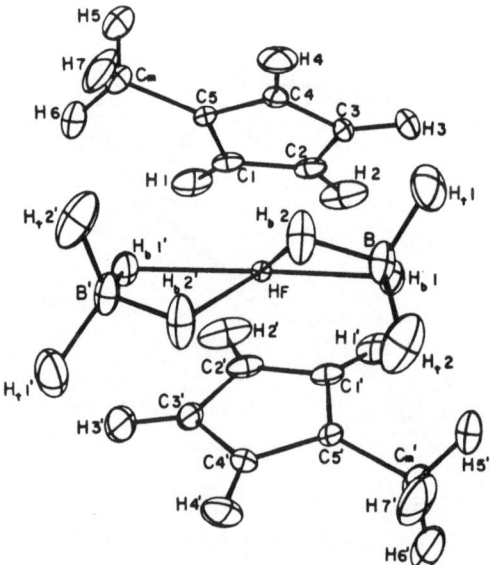

Fig. 18. The structure of $Hf(C_5H_4Me)_2$-$(BH_4)_2$, as determined by neutron diffraction (Ref. 28)

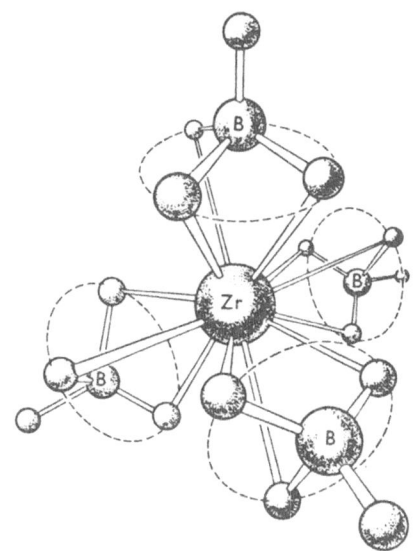

Fig. 19. The structure of $Zr(BH_4)_4$, showing the 12-coordinate zirconium atom. The bridging H atoms were not actually located in this early X-ray study (Ref. 99), but a subsequent neutron diffraction investigation on the isostructural $Hf(BH_4)_4$ confirmed the correctness of this model (Ref. 101)

the first time in an unpublished piece of work by a consortium of investigators[101a], and very recently repeated with higher precision by Marks, Williams and co-workers[101b]. Hafnium borohydride has the same basic structure as zirconium borohydride, and relevant distances and angles are: Hf–H = 2.13(1) Å, B–H = 1.23(1) Å, Hf–B = 2.28(1) Å, Hf–H–B = 80.6(6)°.

Finally, there are complexes in which both bidentate and tridentate BH_4 groups are found. These are represented by the borohydrides of scandium[86a], yttrium[86b] and uranium[87–89,102]. Uranium borohydride in crystalline form has a complicated polymeric structure[102]: each U atom is coordinated to four bidentate and two tridentate $[BH_4]^-$ ligands, resulting in a 14-coordinate complex. The bidentate BH_4 groups form cross-links between U atoms, $U(\mu\text{-H})_2B(\mu\text{-H})_2U$, creating the polymeric network. There is no appreciable difference between the two types of bonding (bidentate vs. tridentate) as far as U–H or B–H bond lengths are concerned, but the U–B distances and U–H–B angles are very different: U–B = 2.86(2) Å, U–H–B = 98(1)° for $U(\mu\text{-H})_2BH_2$; U–B = 2.34(1) Å, U–H–B = 83(1)° for $U(\mu\text{-H})_3BH$. This difference in U–B distance was used to characterize the bonding in a series of ether adducts of uranium borohydride, which can be monomeric, $U(BH_4)_4(THF)_2$ [87]; dimeric, $[U(BH_4)_4(Pr_2^iO)]_2$ [88]; or polymeric, $[U(BH_4)_4(Me_2O)]_n$ [89], $[U(BH_4)_4(Et_2O)]_n$ [89]. In each case, $U(\mu\text{-H})_3BH$ linkages were identified by short U–B distances (2.52–2.56 Å), and $U(\mu\text{-H})_2BH_2$ bonds by long U–B distances (2.87–2.89 Å). These correlations were also found in the structure determinations of $Sc(BH_4)_3(THF)_2$ [86a] and $Y(BH_4)_3(THF)_3$ [86b].

From the limited amount of accurate data available, one can say that M–H–B angles generally decrease in the order $M(\mu\text{-H})BH_3$, $M(\mu\text{-H})_2BH_2$, $M(\mu\text{-H})_3BH$: e.g., M–H–B = 121.7(4)° in $Cu(PPh_2Me)_3(BH_4)$, 96.8(5)° in $Hf(C_5H_4Me)_2(BH_4)_2$, 80.6(6)° in $Hf(BH_4)_4$. And it can be shown that M–B distances also decrease in the above sequence, provided that differences in covalent radii are taken into account.

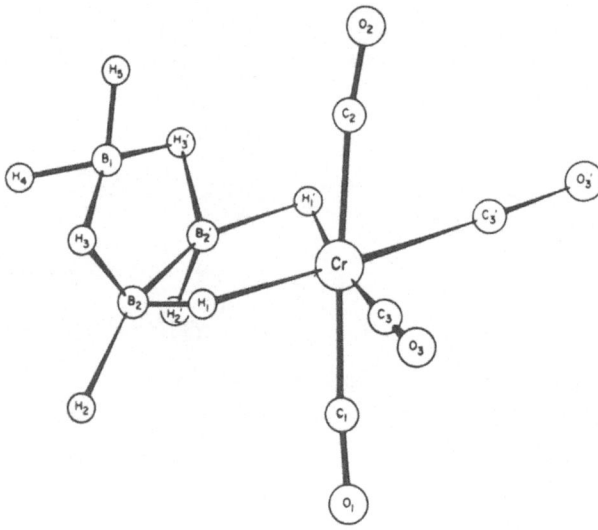

Fig. 20. The familiar bidentate mode of attachment of the $(B_3H_8)^-$ ligand, as found in $[Cr(CO)_4\text{-}(B_3H_8)]^-$ (Ref. 104)

2. Complexes of the Larger Boron Hydrides

In $Cu(PPh_3)_2(B_3H_8)$ [103)] and $[Cr(CO)_4(B_3H_8)]^-$ [104)], the $[B_3H_8]^-$ group was found to be a bidentate ligand (Fig. 20). However, in $Mn(CO)_3(B_3H_8)$ [105)] the $[B_3H_8]^-$ ligand becomes tridentate (Fig. 21). This interesting pattern is repeated in $Mn(CO)_3(B_8H_{12})$ [106)], in which an entire B_3H_3 face of the B_8 ligand participates in a 3-point attachment to the manganese atom.

There are many known compounds in which a transition metal atom becomes incorporated into a borane skeleton [107)] and in some of these cases M–H–B bridges

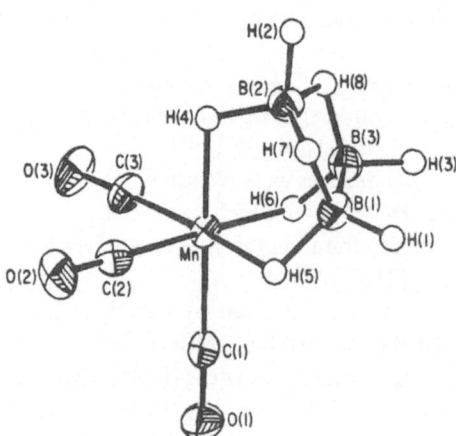

Fig. 21. The unique tridentate mode of binding of the $(B_3H_8)^-$ ligand, as found in $Mn(CO)_3(B_3H_8)$ (Ref. 105). One B–H bond from each boron atom becomes part of a B–H–M bridge to the $Mn(CO)_3$ unit

constitute an integral part of the bonding in the resulting cluster. In this review we will restrict our discussion only to those latter cases. Examples of such molecules include $Mn(CO)_3[B_9H_{12}(THF)]$ [108] and $Co(C_5H_5)(B_9H_{13})$[109], in which a BH unit of $B_{10}H_{14}$ is replaced by a $Mn(CO)_3$ or $Co(C_5H_5)$ fragment. Other selected examples are $[Fe(CO)_3(B_5H_8)]^-$ [110] and $Ir(CO)(PPh_3)_2(B_5H_8)$ [111], which are based on B_6H_{10}, and $Co(C_5H_5)(B_4H_8)$ [112], which is based on B_5H_9.

Two other compounds are worthy of mention. $Cu(PPh_3)_2(B_5H_8)Fe(CO)_3$ [113] (Fig. 22) represents a metalloborane complex having two different metal atoms. The structure is again based on B_6H_{10}: the $Fe(CO)_3$ fragment occupies a basal position normally held by a BH unit, while the $Cu(PPh_3)_2$ moiety has inserted itself into a B−H−B bridge bond. The structure of $Co_2(C_5H_5)_2(B_4H_6)$ [114] is noteworthy in that H atoms were found bridging the Co_2B triangular faces of the octahedral Co_2B_4 cage. We will refer to this and related molecules again in the discussion of face-bridging H atoms in metal clusters (Sect. E. II.).

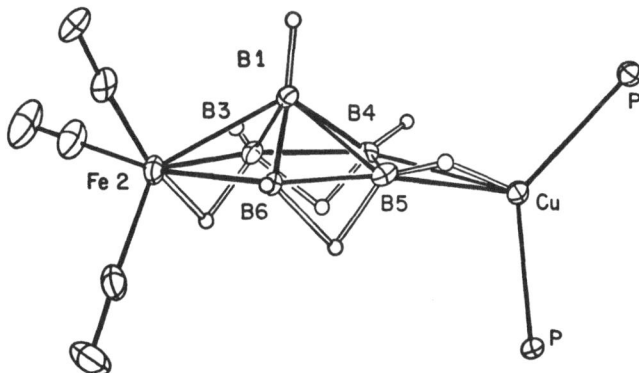

Fig. 22. The structure of the unusual mixed metal boron hydride complex $Fe(CO)_3(B_5H_8)Cu$-$(PPh_3)_2$, whose FeB_5 cage is based on B_6H_{10} (Ref. 113)

As seen earlier in the structures of various $[B_3H_8]^-$ complexes, it often happens that terminal B−H bonds of a boron hydride become converted to B−H−M bridges upon complexation to a metal atom. This phenomenon is frequently found in complexes of closo boranes and carboranes, which only have terminal B−H bonds. Examples include $Cu_2(PPh_3)_4(B_{10}H_{10})$ [115] (Fig. 23), $Cu_2(B_{10}H_{10})$ [116], $Co[(B_{10}C_2H_{10})_2]_2^-$ [117], $Rh(PPh_3)_2(B_{10}C_2H_{10}Ph)$ [118], $[Rh(PPh_3)(B_9C_2H_{11})]_2$ [119] (Fig. 24) and $[Ag(PPh_3)(B_8C_2H_{11})]_2$ [120]. The latter compound contains, to our knowledge, the only example of a silver-hydrogen bond structurally characterized.

A unique metalloborane structure is found in $HMn_3(CO)_{10}(BH_3)_2$ [121] (Fig. 25). Here a B_2H_6 fragment is embedded in the cluster and actually serves to separate the Mn_3 unit into two parts. The cluster is held together by one Mn−H−Mn and six Mn−H−B bridges. A particularly noteworthy feature of this compound is the fact that the B_2H_6 unit which is trapped within it has the otherwise-unknown ethane-like structure.

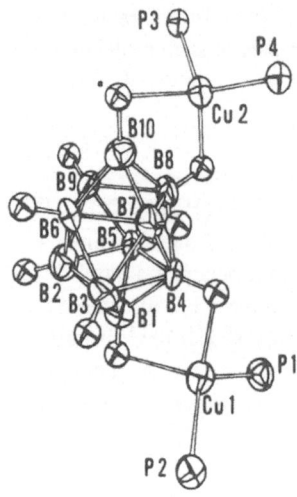

Fig. 23. An illustration of the *closo* $B_{10}H_{10}$ cage serving as a ligand to two $Cu(PPh_3)_2$ units in the structure of $Cu_2(PPh_3)_4(B_{10}H_{10})$ (Ref. 115)

Finally, mention should be made of two compounds, $Mo(CO)_2(\eta^3\text{-}C_3H_5)$-$[H_2B(Me_2pz)_2]$ [122] (Fig. 26) and $Mo(CO)_2(\eta^3\text{-}C_7H_7)[H_2B(Me_2pz)_2]$ [123], in which one of the B–H bonds is placed within interaction distance from the Mo atom as a stereochemical consequence of the conformation of the bis(pyrazolyl)-borato ligand. This is reminiscent of a number of similar molecules having C–H ... M interactions that will be discussed in Sect. C. III.

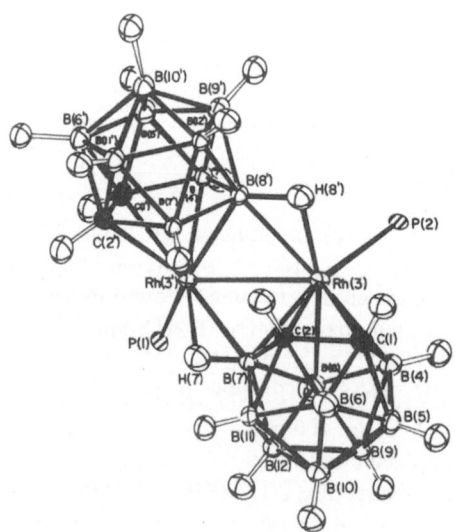

Fig. 24. An ORTEP diagram of $[Rh(PPh_3)$-$(C_2B_9H_{11})]_2$, a molecule with two Rh–H–B bridge bonds. Interestingly, the carbon atoms are asymmetrically positioned with respect to the center of the molecule (Ref. 119)

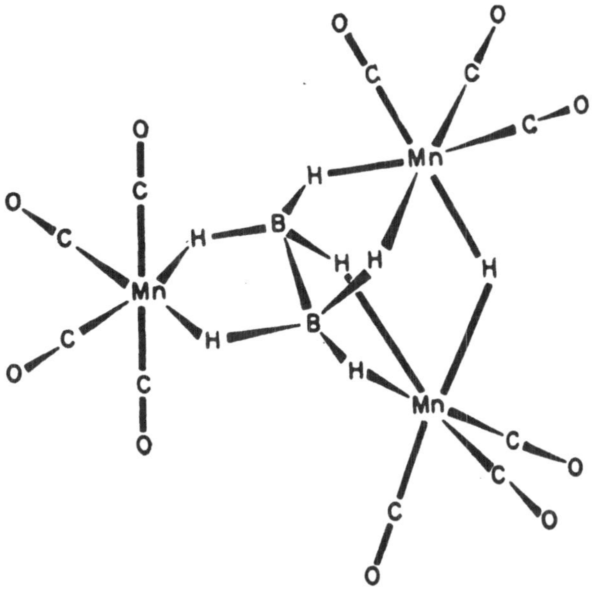

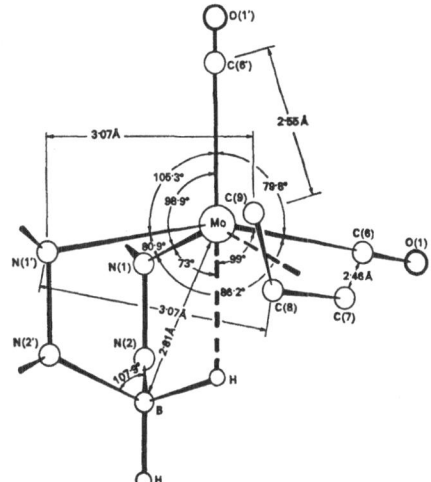

Fig. 25. The structure of $HMn_3(CO)_{10}(BH_3)_2$. Note the ethane-like geometry of the embedded B_2H_6 fragment (Ref. 121)

Fig. 26. The central portion of $Mo(CO)_2$-$(C_3H_5)[H_2B(Me_2pz)_2]$. The conformation of the bis(pyrazolyl)borato ligand places one of the B–H bonds within interaction distance of the molybdenum atom (Ref. 122)

II. M–H–Al Systems

Not much work has been carried out on M–H–Al bonds. The only example in which H atoms have been directly located in such a system is represented by the structure of $HTi_2Cp_2(H_2AlEt_2)(C_{10}H_8)$ [124] (Fig. 27). In this molecule, the molecular parameters associated with the Ti–H–Al bridge are: Ti–H = 1.69 Å, Al–H = 1.70 Å, Ti–Al = 3.13 Å, Ti–H–Al = 135°. Other Ti–H–Al linkages are believed to exist in $[HTi(C_5H_4)AlEt_2]_2(C_{10}H_8)$ [124] and $[HTiCp(C_5H_4)AlEt_2]_2$ [125], and a Mo–H–Al bridge in $[HMoCp(C_5H_4)]_2Al_3Me_5$ [126]. An alternative arrangement is found in $\{Ta[H_2Al(OC_2H_4OMe)_2](Me_2PCH_2CH_2PMe_2)_2\}_2$ [127], in which the molecule is held together by what are believed to be $Ta(\mu\text{-}H)_2Al$ linkages.

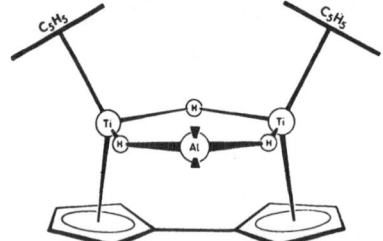

Fig. 27. A schematic drawing of the structure of $HTi_2Cp_2(H_2AlEt_2)(C_{10}H_8)$, showing the presence of a Ti–H–Al bridge bond (Ref. 124)

III. M–H–C Systems

Because of the possibility that they might exist as intermediates in catalytic reactions, compounds with M–H–C interactions constitute perhaps the most important class of molecules of the M–H–X type. However, it was not until recently that molecules having such interactions were isolated and structurally characterized. As an example of a process in which M–H–C interactions are involved, consider metal insertion into a C–H bond. For this process one could envisage the following sequence of events:

$$
\left\{
\begin{array}{c}
C\text{—}H\cdots M \\
\text{or} \\
H \\
| \\
\underset{C}{|}\text{----}M
\end{array}
\right\}
\longrightarrow
\underset{C}{\overset{H}{\diagdown}}\,M
\longrightarrow
\overset{H}{\underset{}{C\text{—}M}}
$$

$$VI \qquad\qquad VII \qquad\qquad VIII$$

In (*VI*), an "end-on" or "side-on" approach of the metal atom to the C–H bond is depicted. This weak interaction between the metal atom (presumably unsaturated) and the C–H bond could conceivably lead to a three-center-bonded intermediate (*VII*), which would in turn transform into the final insertion product (*VIII*). Although many molecules of type *VIII* are known in the literature [128], species that model *VI* and *VII* are rare. The problem is that the process *VI–VII–VIII* will occur spontaneously, unless steric constraints are imposed to prevent the C–H unit from approaching the metal atom too closely. Only then can the process be "arrested" at the *VI* or *VII* stage.

Perhaps the first paper in which the potential existence of a M $\cdots$ H–C interaction was explicitly noted was the report of $Pd(PPh_3)_2[C_4(COOMe)_4H](Br)$ by Roe, Maitlis and co-workers in 1972 [129]. In this structure determination (Fig. 28), the estimated H position corresponded to a Pd $\cdots$ H distance of 2.3 Å, which was considered well below the expected van der Waals' contact distance between Pd and H.

The general problem of M–H–C interactions was brought into spotlight and analyzed in depth in a series of papers by Cotton and co-workers [123,130,131]. The molecules they studied were several interesting pyrazolyl complexes first synthesized by Trofimenko [132]. In an earlier section of this article, we drew attention to two

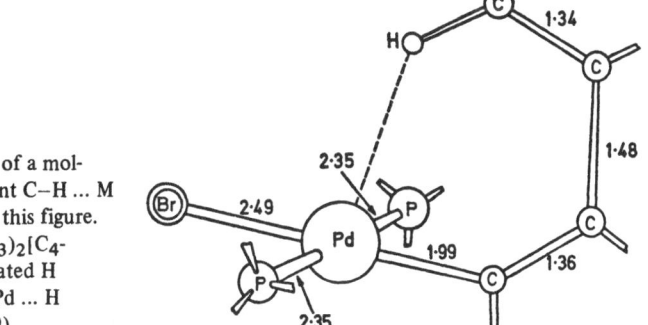

Fig. 28. An early example of a molecule containing an incipient C–H ... M interaction is illustrated in this figure. In the structure of $Pd(PPh_3)_2[C_4-(COOMe)_4H]Br$, the estimated H position corresponds to a Pd ... H distance of 2.3 Å (Ref. 129)

pyrazolyl complexes of the type $Mo(CO)_2(\eta^3\text{-allyl})[H_2B(pz)_2]$, in which strong B–H $\cdots$ M interactions were found (Fig. 26). Recognizing that the replacement of a hydrogen atom by an alkyl group might give rise to compounds having C–H $\cdots$ M interactions, Cotton and co-workers studied the ethyl analogues $Mo(CO)_2$-$[\eta^3\text{-}CH_2C(Ph)CH_2][Et_2B(pz)_2]$ [130] and $Mo(CO)_2[\eta^3\text{-}C_7H_7][Et_2B(pz)_2]$ [131]. The result for the former compound is shown in Fig. 29. The refined H position, H(6), was found to be 2.27(8) Å from the Mo atom, a distance corresponding to a weak but significant Mo $\cdots$ H interaction. The chemical inertness of this formally 16-electron complex can be rationalized by proposing that the Mo atom is receiving additional electron density from the C–H bond, and is therefore "pseudo-saturated". Incidentally, a good summary of the background literature on molecules having suspected M $\cdots$ H–C interactions is given in Ref. 130.

The most definitive example of C–H $\cdots$ M interaction is represented by the neutron diffraction analysis of $\{Fe[P(OMe)_3]_3(\eta^3\text{-}C_8H_{13})\}[BF_4]$ by Williams and co-workers [133]. In this structure, the CH_2 group adjacent to the π-allyl portion of

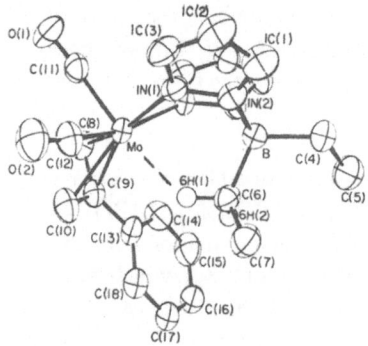

Fig. 29. A molecular plot of $Mo(CO)_2$-$(C_4H_3Ph)[Et_2B(pz)_2]$, showing the suspected Mo ... H–C interaction (*dotted line*). The hydrogen atom indicated in this diagram is estimated to be 2.27 Å from the Mo atom. (Ref. 130)

the C_8H_{13} ring has one of its H atoms situated 1.874(3) Å from the iron atom (Fig. 30). This distance represents the shortest known M–H separation in a M $\cdots$ H–C system, and is in fact only about 0.1 Å longer than a normal covalent Fe–H (bridging) distance. In addition, the C–H bond which is interacting with the Fe atom is significantly longer than usual [1.164(3) Å], a fact which is again consistent with the presence of a strong M $\cdots$ H–C interaction.

Other, less well-defined examples of suspected M $\cdots$ H–C interactions have appeared in the literature [134–147]. One is illustrated by the structure of the T-shaped $[Rh(PPh_3)_3]^+$ cation [140]. The rhodium atom in this formally 14-electron complex is apparently trying to compensate for its extreme electron deficiency by interacting with the C–C–H portion of one of its nine phenyl rings (Fig. 31). This results in one of the Rh–P–C angles being severely distorted [75.6(5)°] from a normal tetrahedral value. In this case, the Rh $\cdots$ H–C interaction is augmented by

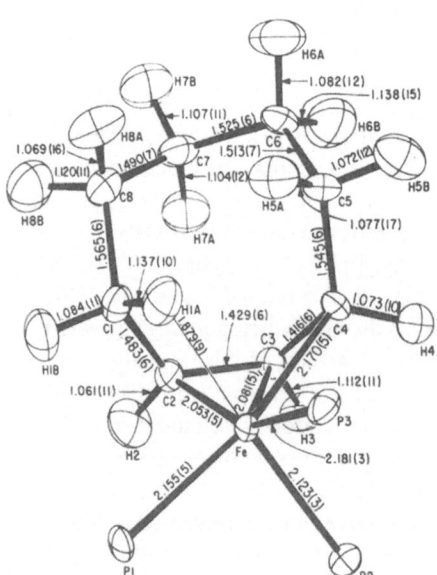

Fig. 30. The results of a neutron diffraction analysis of the $\{Fe[P(OMe)_3]_3(\eta^3\text{-}C_8H_{13})\}^+$ cation, which provided the first definitive evidence for metal-hydrogen-carbon interaction (Ref. 133)

Fig. 31. A close-up of the interaction region between the rhodium atom and the unique phenyl group in the [Rh-(PPh$_3$)$_3$]$^+$ cation (Ref. 140). There are weak but perceptible interactions between Rh and the C$_1$ and C$_2$ atoms, causing the Rh−P$_1$−C$_1$ angle to assume a highly distorted value of 75.6 (5)°. The ortho hydrogen atom, whose calculated position is indicated in this diagram, also appears to be weakly interacting with the metal

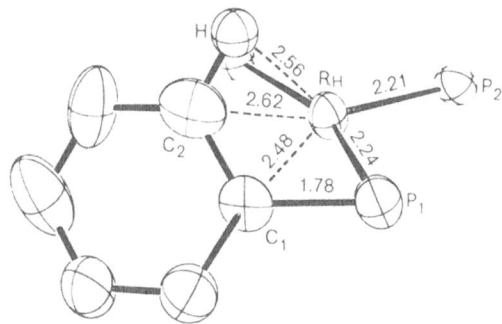

a Rh ⋯ alkene interaction, and the complex as a whole is suggestive of an ortho-metalation reaction [128] about to take place.

One other type of M ⋯ H−C interaction in a mononuclear complex is worthy of mention. In the neutron diffraction study of [Ta(=CHCMe$_3$)(PMe$_3$)Cl$_3$]$_2$ [148], the alkylidene H atom [H(1) in Fig. 32] is significantly distorted towards the tantalum atom. This is shown by the fact that the Ta−C(1)−H(1) angle (84.8°) is much smaller than the C(2)−C(1)−H(1) angle (113.7°), and by the fact that the C(1)−H(1) distance is, as in the {Fe[P(OMe)$_3$]$_3$(η^3-C$_8$H$_{13}$)}$^+$ complex discussed earlier, significantly longer than usual [1.131(3) Å]. The Ta ⋯ H distance in this compound, 2.119(4) Å, is only about 0.15 Å longer than that expected for a normal Ta−H (bridging) distance. Similar features are found in other metal complexes of bulky alkylidenes, e.g., Ta(=CHCMe$_3$)$_2$ (mesityl)(PMe$_3$)$_2$ [149] and W(≡CCMe$_3$)-(=CHCMe$_3$)(CH$_2$CMe$_3$)(dmpe) [150], which have been analyzed with X-ray diffraction. In these compounds, the M=C−C angles (150°−170°) are very much distorted from values expected for normal sp^2 carbon atoms (i.e., 120°), indicating again that some sort of M ⋯ H−C interaction is almost certainly taking place.

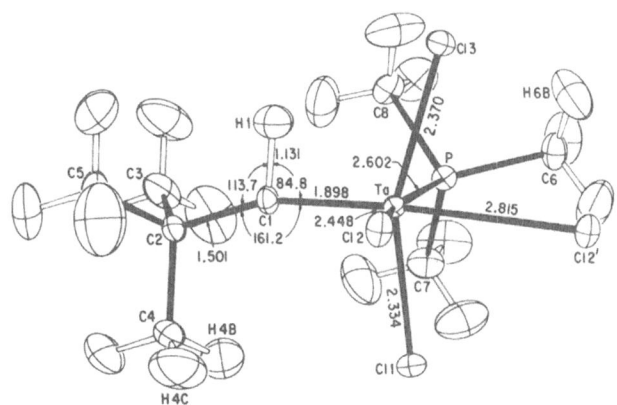

Fig. 32. The results of a neutron diffraction analysis on [Ta(=CHCMe$_3$)(PMe$_3$)Cl$_3$]$_2$, showing the unexpectedly acute Ta−C$_1$−H$_1$ angle which indicates the presence of some Ta ... H interaction. (Ref. 148)

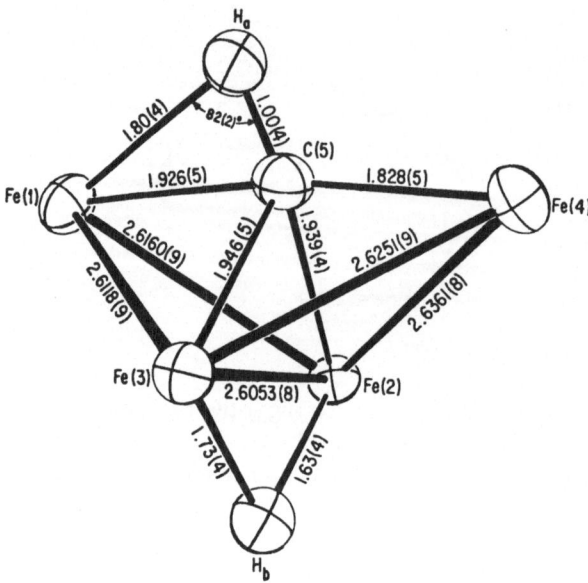

Fig. 33. A plot of the core of the $HFe_4(\eta^2\text{-CH})(CO)_{12}$ molecule, showing the unprecedented feature of a η^2-CH group embedded in the cluster (Ref. 151). Note that the C–H group is interacting with Fe(1) in a "sideways" fashion. The X-ray derived distances and angles shown in this figure (Ref. 151a) have been subsequently updated by more accurate neutron results (Ref. 151b). The gross structure of this molecule is very similar to that of $[HFe_4(CO)_{13}]^-$ (Fig. 49)

Finally, in a very recent X-ray and neutron study of the cluster complex $HFe_4(\eta^2\text{-CH})(CO)_{12}$ [151)] (Fig. 33) an "open-butterfly" geometry is found, very similar to that of the well-known $[HFe_4(CO)_{13}]^-$ [152)]. The two compounds can be considered isoelectronic, with the $[\equiv C–H]$ group in one corresponding to the $[\equiv C–O^-]$ unit in the other. Interest in $HFe_4(\eta^2\text{-CH})(CO)_{12}$ stems from the unprecedented feature of a η^2-CH unit embedded in a metal cluster. The carbon atom of the CH unit is coordinated to all four iron atoms, and the side-on bonding of CH to Fe is suggestive of structure *VI* mentioned at the beginning of this section. The Fe–H distance in this molecule is 1.750(4) Å. The side-on nature of the (C–H) ··· M binding found here (*VI*) may reflect the preferred orientation of a C–H fragment on a metal surface.

It should be added that there are other examples of cluster complexes in which C–H ··· M bonding is believed to take place (e.g., the CH_3 ligand in $HOs_3(CO)_{10}(CH_3)$ (*IX*) [153)]), but these have not yet been characterized structurally.

IV. M—H—Si Systems

Compounds having M—H—Si bridges are of interest because they may be considered models for M—H—C interactions. Structural work done in this area has largely been carried out on compounds synthesized by Graham and co-workers, and is nicely summarized in a paper by Cowie and Bennett [154].

The first M—H—Si system investigated was the structure of $Re_2(CO)_8(H_2SiPh_2)$, reported by Elder in 1970 [155]. The H atoms were not located, but were assumed to bridge the Re—Si edges of the Re_2Si triangle (X). It was pointed out that the estimated H positions corresponded to reasonable Re—H and Si—H distances. However, this conclusion was later challenged by Bennett and co-workers [154,156,157], who felt that the H ligand should more properly be considered terminal (XI). Their argument was based largely on the fact that the Re—Si distances in $Re_2(CO)_8(H_2SiPh_2)$ are essentially the same as those in complexes that contain only unbridged Re—Si bonds (XII), e.g., $Re_2(CO)_8(SiPh_2)_2$.

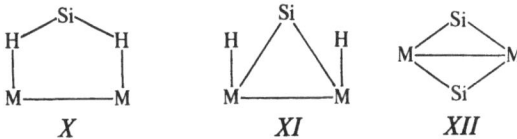

The structure of $W_2(CO)_8H_2(SiPh_2)_2$ [156] (Fig. 34) now appears to be the only metal carbonyl in which *bona fide* M—H—Si bridges are believed to exist. Again, the H atoms were not located, but there are distinct clues which indicate the presence of H bridges. The W_2Si_2 core of the molecule exhibits a characteristic two-short/two-long pattern of W—Si distances (2.586, 2.703 Å) which correspond to W—Si and W—H—Si bonds respectively (XIII). Moreover, the skewed orientation of the car-

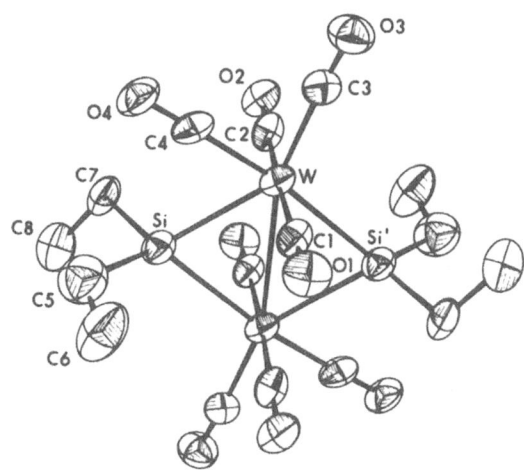

Fig. 34. An ORTEP diagram of $H_2W_2(CO)_8(SiEt_2)_2$. The presence of two long and two short W—Si bonds in this dimer strongly suggests that the hydrogen atoms are bridging the two longer bonds (labelled W—Si'). This conclusion is also supported by the fact that the C_3—W—Si' angle is much wider than the C_4—W—Si angle (Ref. 156)

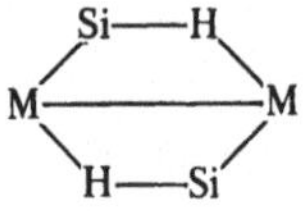

Si—H
M————M
H—Si

XIII

bonyl groups (C_4–W–Si = 66.6°, C_3–W–Si' = 109.1°; see Fig. 34) is also consistent with H atoms bridging the long W–Si' edges.

In $Re_2(CO)_7H_2(SiEt_2)_2$ [157)] and $Re_2(CO)_6H_4(SiEt_2)_2$ [154)], the H atoms are believed to be terminal (*XIV* and *XV* respectively). $Re_2(CO)_7H_2(SiEt_2)_2$ provides a good test case since, if the H atoms had been bridging, one would have seen a two-long/two-short pattern of M–Si distances, as was found in $W_2(CO)_8H_2(SiPh_2)_2$ (*XIII*). However, the four Re–Si bond lengths in $Re_2(CO)_7H_2(SiEt_2)_2$ are essentially identical (*XIV*), and are moreover indistinguishable from those in $Re_2(CO)_6H_4(SiEt_2)_2$ (*XV*), $Re_2(CO)_8(H_2SiPh_2)$ (*XI*) and $Re_2(CO)_8(SiPh_2)_2$ (*XII*). Hence it was concluded that the Re–Si bonds in all these compounds are unbridged [154,157)]. In the case of $Re_2(CO)_6H_4(SiEt_2)_2$ (*XV*), the H atoms were actually found from a difference Fourier map, but could not be refined. Their positions were such that one could argue that weak Si ⋯ H interactions might be present to augment the terminal M–H bonds (*XVI*). It was not possible, from the available evidence, to determine unambiguously if such interactions are present or not.

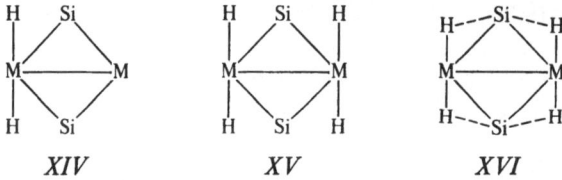

| *XIV* | *XV* | *XVI* |

In two recent publications by Stone, Spencer, Howard and co-workers, the structures of $[Pt(\mu\text{-}H)(\mu\text{-}SiMe_2)(PCy_3)]_2$ [158)] and $[Pt(\mu\text{-}H)(SiEt_3)(PCy_3)]_2$ [159)] were described. In $[Pt(\mu\text{-}H)(\mu\text{-}SiMe_2)(PCy_3)]_2$ the H atoms were located in positions corresponding to structure *XVII*, 1.78 Å from Pt and 1.72 Å from Si. In $[Pt(\mu\text{-}H)(SiEt_3)$-$(PCy_3)]_2$ an unusual form of multicenter (Si ⋯ H ⋯ Pt)Pt interaction is proposed (*XVIII*), a conclusion which is supported by an anomalous low-field chemical shift (τ 6.2 ppm) of the bridging H atoms (which indicates some Si ⋯ H interaction), and an anomalous ν(Pt–H) IR band at 1655 cm^{-1} (too high in frequency to be a Pt–H–Pt band), which is also suggestive of delocalized 3 c–2 e Pt–H–Si bonding.

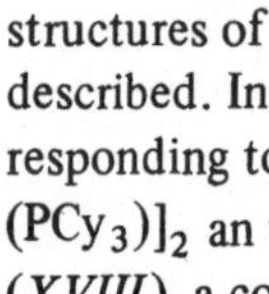

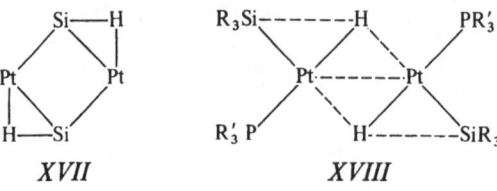

| *XVII* | *XVIII* |

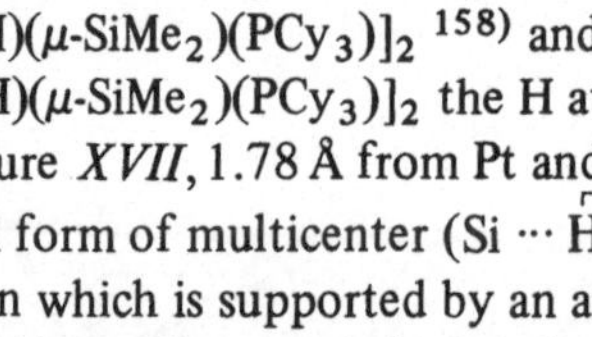

However, very preliminary neutron results on the related $\{Pt(\mu\text{-}H)[Si(OEt)_3]_2\text{-}(PMeBu_2^t)\}_2$ show a normal $Pt(\mu\text{-}H)_2Pt$ linkage with no apparent H ⋯ Si interaction [160].

Suspected Si ⋯ H—M interactions were also discussed in connection with the mononuclear complexes $HReCp(CO)_2(SiPh_3)$ [161], $HMnCp(CO)_2(SiPh_3)$ [161,162] and $HFeCp(CO)_2(SiF_2Me)_2$ [163]. From an analysis of known or estimated Si ⋯ H distances, it was concluded that Si ⋯ H interactions were most likely absent in the rhenium and iron complexes. In the case of $HMnCp(CO)_2(SiPh_3)$, it was originally believed that a true example of Mn—H ⋯ Si interaction existed [162], but a subsequent re-assessment of the problem indicates that the structural evidence is, at best, inconclusive [161,163].

It should be pointed out that systems with M—H ⋯ Si interactions investigated so far differ fundamentally from the M ⋯ H—C systems discussed in the previous section. In the former case, the main issue is whether Si ⋯ H interactions are present or not, while in the latter case it is the M ⋯ H portion which is in question.

V. Other Systems: M–H–Zn and M–Li–M

To our knowledge, the only example of a M—H—Zn bridge involving a transition metal is represented by the complex $H_2MoCp_2 \cdot ZnBr_2 \cdot DMF$ (Fig. 35) [164]. In this compound, weak Zn ... H interactions are believed to exist as part of a bridging $Mo(\mu\text{-}H)_2Zn$ fragment. The H atoms were located from a difference Fourier map but were not refined. The observed Zn—H (bridging) distance of 1.69 Å found in this complex compares very well with the Zn—H (terminal) distance of 1.62 Å in $[HZnN(Me)C_2H_4NMe_2]_2$, derived from neutron diffraction data [165].

Interest in the tetrameric molecules $[HMoCp_2Li]_4$ [166,167] and $[HWCp_2Li]_4$ [167] stems not from their metal-hydrogen linkages (which are simply terminal) but from the unprecedented metal-lithium-metal bridges found in them (Fig. 36). These molecules, whose structures were reported by Green, Prout and co-workers, contain discrete eight-membered M_4Li_4 rings with the following distances and angles:

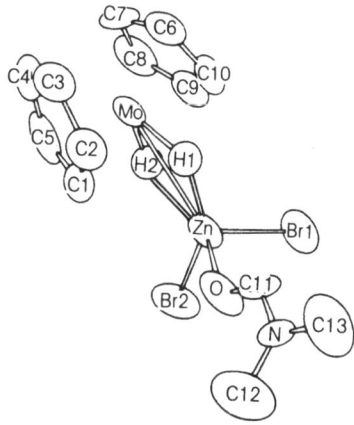

Fig. 35. A very rare example of a hydrogen bridge bond involving a zinc atom, as found in the structure of $H_2MoCp_2 \cdot ZnBr_2 \cdot DMF$ (Ref. 164). The average Zn—H distance of 1.69 Å found in this complex is, in fact, that expected for a Zn—H bridge, based on the known Zn—H (terminal) distance of 1.62 Å in $[HZnN(Me)\text{-}C_2H_4NMe_2]_2$ (Ref. 165)

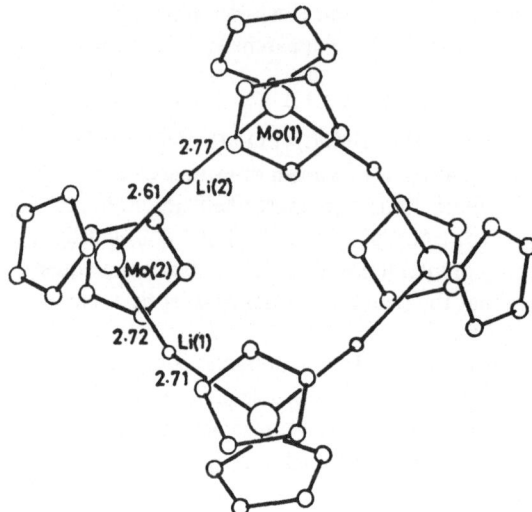

Fig. 36. An illustration of the highly unusual metal-lithium-metal bridge, as found in [HMoCp$_2$Li]$_4$ (Ref. 166). The Mo—Li—Mo angles in this complex (165°) are much more obtuse than those of typical M—H—M bridges. Terminal H atoms (not located) are believed to bisect the Li—Mo—Li angles

Mo—Li = 2.70 Å, Li—Mo—Li = 99°, Mo—Li—Mo = 165°; W—Li = 2.69 Å, Li—W—Li = 98°, W—Li—W = 166°. The terminal H atoms (not located) are believed to bisect the Li—M—Li angles. The possible existence of H—M—Li 3-center interactions was speculated upon [166]. The M—Li—M bridges found here are reminiscent of the well-studied M—H—M bridge bonds (to be discussed in the next section), differing mainly in that the bridging atom uses a 2 s orbital (of Li) as opposed to a 1s orbital (of H).

D. M—H—M Bonds

In the past several years, M—H—M bridge bonds have been intensively studied with neutron diffraction techniques, in bimetallic complexes as well as in metal clusters. In every system investigated so far, the M—H—M linkage has been found to be distinctly bent, with M—H—M angles ranging from 85° to 159° (most commonly, in the range 100° to 130°). This situation (i.e., the non-linearity of the H bridge) may prevail for most other types of 3-center/2-electron X—H—X bonds[4].

I. M(μ-H)M Systems

HW$_2$(CO)$_9$(NO) [23] (Fig. 37) and HW$_2$(CO)$_8$(NO)[P(OMe)$_3$] [168] were the first molecules with unsupported^c M—H—M bonds to have been investigated with single-

---

^{c} We define a molecule having an unsupported M—H—M bond as one which is held together solely by that bond. Thus, molecules having other bridging ligands, such as M(μ-H)(μ-X)M, M(μ-H)$_2$M or M(μ-H)M systems, are excluded by this definition

$\underset{\rule{1.2em}{0pt}\text{M}\rule{1.2em}{0pt}}{\underbrace{\rule{2.4em}{0pt}}}$

Fig. 37. The structure of
$HW_2(CO)_9(NO)$ as determined
by neutron diffraction (Ref.
23a). Note that the axial ligand-
metal vectors do not point at
the bridging H atom, but at
the center of the W—H—W
triangle. This observation was
taken as evidence that there is
significant metal-metal overlap
in a M—H—M 3-center-2-electron
bond. The structure shown here
illustrates one advantage of
using metal complexes to study
electron-deficient bonding:
the ML_5 moiety on each metal
atom serves as a convenient
coordinate system to pinpoint
the direction of the orbital
used by tungsten to participate
in W—H—W overlap

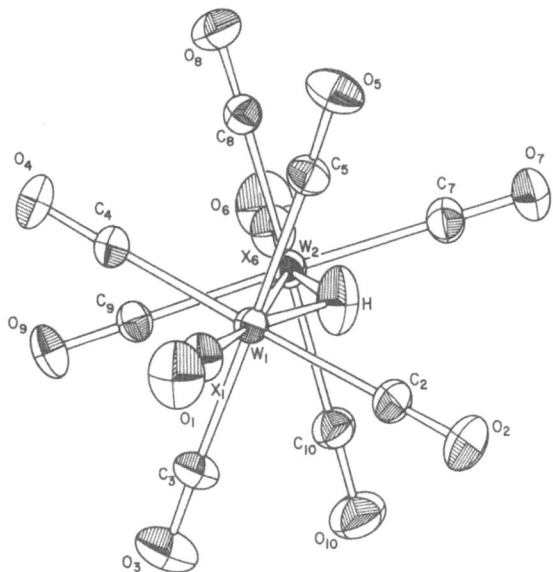

crystal neutron diffraction techniques. Their structures have been described in earlier reviews[4] and will not be repeated in depth in this article. Suffice it to say that the H atoms in these molecules are located in characteristically bent, off-axis positions (*XIX*), which is consistent with the presence of closed 3-center/2-electron bonding (*XX*) (i.e., one in which all 3 orbitals overlap in a common region of space).

XIX   *XX*

The literature on the structures of the $[HM_2(CO)_{10}]^-$ ions (M = Cr, Mo, W) is extensive. The $[HCr_2(CO)_{10}]^-$ ion was originally believed to have a linear, symmetric Cr—H—Cr core, based on X-ray work carried out on its $[NEt_4]^+$ salt [169]. Later neutron diffraction studies [170], however, showed the Cr—H—Cr bond to be bent [158.9(6)°], with the bridging H atom located in a two-fold disordered position (Fig. 38). This work was extended to the $[(Ph_3P)_2N]^+$ salt of the same anion: neutron diffraction investigations on $[(Ph_3P)_2N]^+[HCr_2(CO)_{10}]^-$ [171] revealed what appeared to be a linear M—H—M bridge. However, the thermal ellipsoid of the bridging H atom had a highly distorted plate-shaped appearance, which indicated that the H atom is actually situated off the Cr—Cr axis, probably in a four-fold disordered manner. Subsequent work on the deuterated analogue, $[(Ph_3P)_2N]^+$-

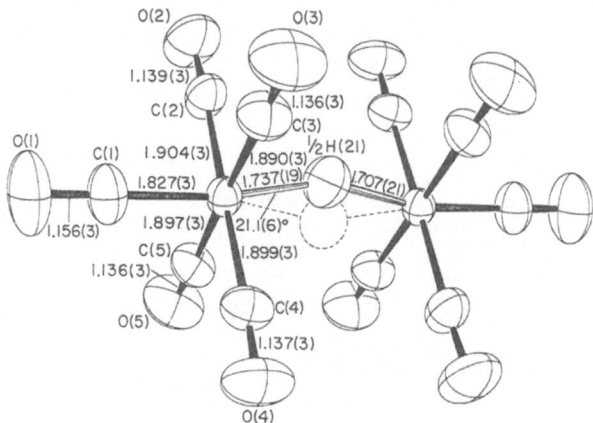

Fig. 38. The structure of the $[HCr_2(CO)_{10}]^-$ anion in $[NEt_4]^+[HCr_2(CO)_{10}]^-$ based on neutron diffraction data (Ref. 170). In contrast to an earlier conclusion (Ref. 169), the hydrogen atom is seen to be off-axis (0.3 Å from the center of the Cr–Cr bond), giving rise to a bent Cr–H–Cr linkage [158.9 (6)°]. A crystallographic center of inversion exists at the center of the anion, giving rise to a disordered superposition of two bent Cr–H–Cr linkages

$[DCr_2(CO)_{10}]^-$ [172], confirmed the four-fold disordered, bent M–H–M model [Cr–D–Cr = 156(1)°].

Initial X-ray studies on the analogous tungsten ion, $[HW_2(CO)_{10}]^-$, showed that the non-hydrogen skeleton could adopt different geometries depending on the counter-ion used [173]. $[NEt_4]^+[HW_2(CO)_{10}]^-$ is isostructural with its Cr analogue and has what appeared to be a linear/eclipsed $W_2(CO)_{10}$ framework, while the $[HW_2(CO)_{10}]^-$ ion in $[(Ph_3P)_2N][HW_2(CO)_{10}]^-$, unlike its Cr analogue, has a bent/ staggered non-hydrogen skeleton. Subsequent neutron work on $[NEt_4]^+[HW_2(CO)_{10}]^-$ [4,174] showed a complicated disordered structure with, again, a bent W–H–W linkage [137(1)°], while the $[HW_2(CO)_{10}]^-$ ion in $[PPh_4]^+[HW_2(CO)_{10}]^-$ [4,174] turned out to have a geometry very similar to that of $HW_2(CO)_9(NO)$.

Of the molybdenum members of the $[HM_2(CO)_{10}]^-$ family, only the structure of $[NEt_4]^+[HMo_2(CO)_9(PPh_3)]^-$ has been published to date [175a]. X-ray studies on this salt reveal a bent/staggered geometry with a markedly asymmetric Mo–H–Mo bridge [Mo–H = 1.68(5), 2.19(6) Å]. Preliminary results on the $[Et_4N]^+$ and $[(Ph_3P)_2N]^+$ salts of $[HMo_2(CO)_{10}]^-$ show a situation very similar to that of the tungsten analogues [175b].

From the results accumulated so far, it seems that the bent/staggered geometry (Fig. 37) is favored for the Mo and W anions, while the linear/eclipsed geometry (here the word "linear" refers to the OC–M ··· M–CO axis and not the M–H–M bond) is the dominant structure for the Cr anions (Fig. 38).

A number of mono-hydrogen bridged complexes which also contain terminal hydride ligands have been prepared. Of these, structure determinations have been carried out on $[(Et_3P)_2(H)Pt(\mu\text{-}H)Pt(Ph)(PEt_3)_2]^+[BPh_4]^-$ [176], $\{H_2Pt_2[Bu_2^tP\text{-}(CH_2)_3PBu_2^t]_2(\mu\text{-}H)\}^+[BPh_4]^-$ [177] and $[(HWCp)_2(\mu\text{-}H)]^+[ClO_4]^-$ [38]. In none of

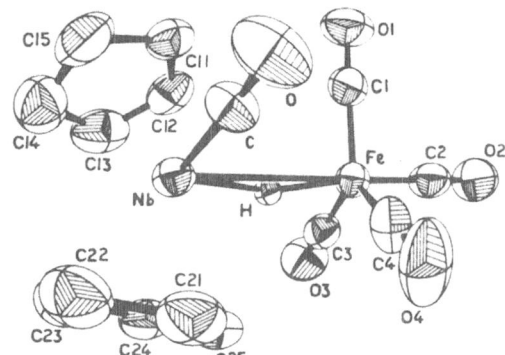

Fig. 39. An illustration of $(OC)_4Fe(H)NbCp_2(CO)$, a rare example of a singly H-bridged species involving two different transition metals (Ref. 178)

these structure determinations were the hydride ligands located, but in the case of $[(HWCp)_2(\mu\text{-}H)]^+$, reasonable estimates could be made by fitting the known structure of H_2MoCp_2 to the non-hydrogen skeleton of the dimeric complex [38].

Finally, the structure of the interesting mixed-metal complex $Cp_2(CO)Nb(\mu\text{-}H)$-$Fe(CO)_4$ (Fig. 39) has recently been analyzed by X-ray diffraction [178]. It shows, as expected, a bent M–H–M' bond with an asymmetrically-positioned H bridge: Nb–H = 1.91(3) Å, Fe–H = 1.61(3) Å, Nb–H–Fe = 141(2)°.

II. $M(\mu\text{-}H)_2M$ Systems

Among the earliest examples of $M(\mu\text{-}H)_2M$ systems structurally investigated are the isoelectronic complexes $H_2Re_2(CO)_8$ [179] and $[H_2W_2(CO)_8]^{2-}$ [180]. The structure

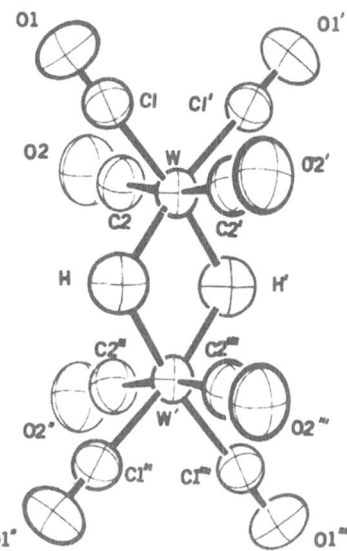

Fig. 40. A molecular plot of the dianion $[H_2W_2(CO)_8]^{2-}$ with the two bridging hydrogen atoms as determined by X-ray diffraction. This structure determination represents one of the few instances in which H atoms attached to third-row transition metals were successfully refined (Ref. 180)

determination of $[H_2W_2(CO)_8]^{2-}$ (Fig. 40) is especially noteworthy since it represents one of the earliest instances in which H atoms directly bonded to *third-row* transition metals were located and refined.

Two other $M(\mu\text{-}H)_2M$ systems analyzed by X-ray diffraction are $\{Ni(\mu\text{-}H)\text{-}[Cy_2P(CH_2)_3PCy_2]\}_2$ [181] (Fig. 41) and $[Pt(\mu\text{-}H)(SiEt_3)(PCy_3)]_2$ [159]. In the former case, one finds the two NiP_2 units tilted 63° relative to each other, with the central $Ni(\mu\text{-}H)_2Ni$ plane bisecting this dihedral angle. Theoretical calculations indicate that the potential energy surface corresponding to this type of twisting deformation is very soft [181].

Neutron diffraction work has been carried out on $H_2Rh_2[P(OPr^i)_3]_4$ [182] and $H_4Th_2(C_5Me_5)_4$ [55]. The initial analysis of $H_2Rh_2[P(OPr^i)_3]_4$ [182a] was of rather low precision owing to small crystal size, but nevertheless an approximately planar $H_2Rh_2P_4$ central fragment could be clearly seen. A subsequent reanalysis produced more precise distances and angles [182c]. $H_4Th_2(C_5Me_5)_4$ (Fig. 42) represents the first organometallic actinide hydride complex structurally characterized. The central $Th(\mu\text{-}H)_2Th$ fragment [dimensions: Th–Th = 4.007(8) Å, Th–H = 2.29(3) Å, Th–H Th = 122(4)°, H–Th–H = 58(1)°] contains a few interesting features, such as an unusually acute H–M–H angle, and an unusually long Th ⋯ Th distance. In contrast to all other $M(\mu\text{-}H)_2M$-bonded complexes described in this section, the M–M distance in this compound is actually longer than expected for a M–M single bond.

The recent structure determination of $[Cp_2W(\mu\text{-}H)_2Rh(PPh_3)_2]^+$ [183] (Fig. 43) represents the first example of a $M(\mu\text{-}H)_2M'$ linkage (i.e., one that involves two different metals) structurally characterized. The H atoms were located and included in the final refinements, but their positions were not reported in the paper.

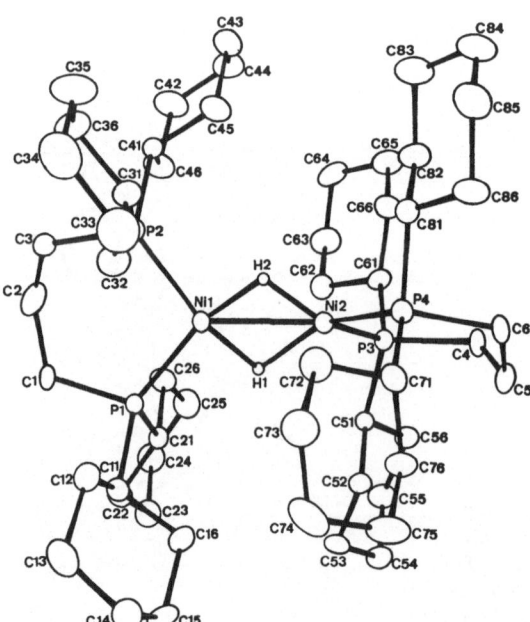

Fig. 41. The structure of $H_2Ni_2[Cy_2P(CH_2)_3PCy_2]_2$ (Ref. 181). The two P_2Ni planes are tilted relative to each other, and the central Ni_2H_2 plane bisects the dihedral angle between them

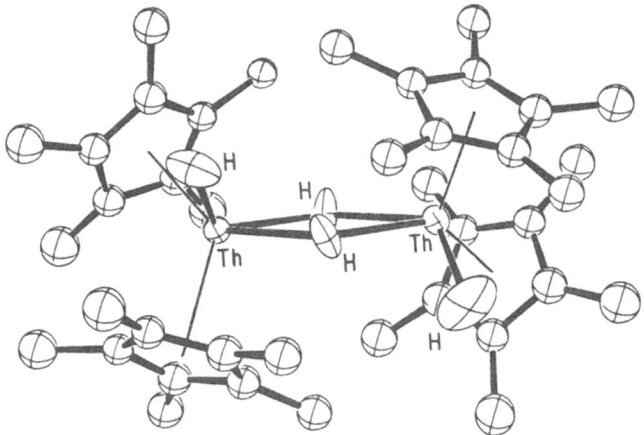

Fig 42. The structure of $H_4Th_2(C_5Me_5)_4$, as determined by neutron diffraction (Ref. 55). The terminal and bridging Th$-$H distances [2.03 (1) and 2.29 (3) Å] are among the longest M$-$H distances known to date

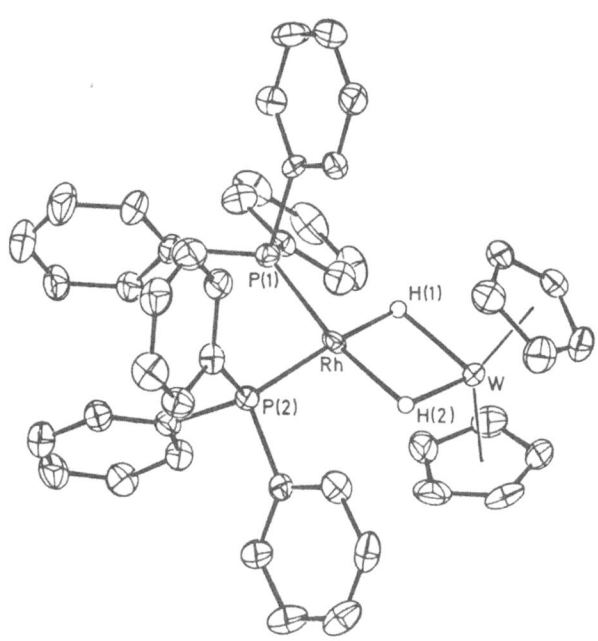

Fig. 43. The only known example of a doubly H-bridged species involving two different transition metals, as illustrated by the structure of $(Ph_3P)_2Rh(\mu\text{-}H)_2WCp_2$ (Ref. 183)

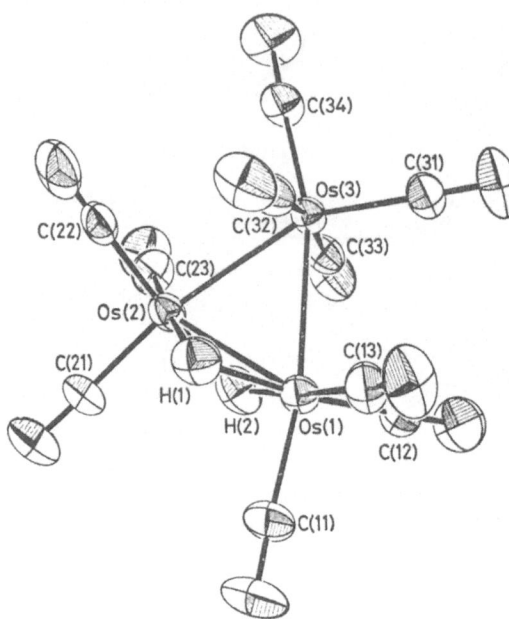

Fig. 44. The structure of $H_2Os_3(CO)_{10}$, which has been analyzed twice by X-ray diffraction (Ref. 203, 204) and twice by neutron diffraction (Ref. 30, 184). This particular diagram, taken from Ref. 30a, represents the result of combined X-ray/neutron least-squares refinement

The $M(\mu\text{-}H)_2M$ unit is sometimes found as part of a metal cluster framework. It has been accurately characterized by neutron diffraction in $H_2Os_3(CO)_{10}$ [30,184] (Fig. 44), and is also suspected to exist in $[H_3Re_3(CO)_{10}]^{2-}$ [185] and $[H_4Re_3(CO)_{10}]^-$ [186]. In contrast to the H atoms in singly-bridged complexes such as $[Ph_4P]^+$-$[HW_2(CO)_{10}]^-$, which are found "outside" the intersection point of the *trans* ligand-metal vectors (*XXI*) [4], the H atoms in $H_2Os_3(CO)_{10}$ are located "inside" the intersection region of the $OC-M$ vectors (*XXII*) [30,184]. In this and other complexes of the $M(\mu\text{-}H)_2M$ type, the bonding in the central region is usually described in terms of four-center interactions [179,182,184].

XXI   *XXII*

III. $M(\mu\text{-}H)_3M$ and $M(\mu\text{-}H)_4M$ Systems

Such systems are rare. The first examples of $M(\mu\text{-}H)_3M$ linkages were found by Dapporto, Midollini and Sacconi in the complexes $\{H_3Fe_2[(Ph_2PCH_2)_3CCH_3]_2\}$-$[PF_6]$ and $\{H_3Co_2[(Ph_2AsCH_2)_3CCH_3]_2\}[BPh_4]$ (Fig. 45) [187]. The H atoms were located in both X-ray structure investigations and corresponded to average

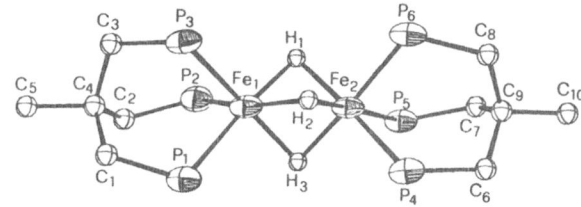

Fig. 45. The first example of an M(μ-H)₃M linkage, as found in the structure of $\{H_3Fe_2[(Ph_2PCH_2)_3CCH_3]_2\}[PF_6]$ (Ref. 187)

Fe−H and Co−H distances of 1.83(3) and 1.70(4) Å respectively. Interestingly, the Fe−Fe and Co−Co distances in these two compounds are almost equal [2.332(3) and 2.377(8) Å respectively] even though the complexes differ in electron configuration (by two electrons).

The complex $[H_3Ir_2(C_5Me_5)_2][BF_4]$ has been analyzed by single-crystal neutron diffraction techniques [51]. The average Ir−H distance was found to be 1.75 Å, but this value may be subject to a rather large uncertainty, since crystal quality was poor and the effects of thermal motion were considerable. The iso electronic complex $[H_2Ir_2(PPh_3)_4(\mu\text{-}H)_3]^+$ has been analyzed by X-ray diffraction [188], and is also believed to contain a M(μ-H)₃M core.

The only M(μ-H)₄M linkage structurally characterized (indeed, the only compound known to contain such a bonding arrangement) is represented by H_8Re_2-$(PEt_2Ph)_4$ [60]. Figure 46 shows very clearly the quartet of H atoms surrounding the Re−Re bond, as revealed by a neutron diffraction study. The bonding in this unusual compound has been rationalized in molecular orbital terms by Dedieu, Albright and Hoffmann [189]. The molecule is surprisingly fluxional: all eight H atoms are equivalent on the NMR time scale even at low temperature [60].

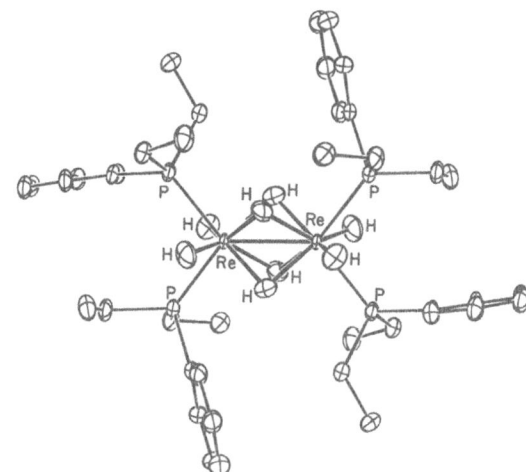

Fig. 46. The unprecedented feature of a quadruply H-bridged metal-metal bond, as found in $H_8Re_2(PEt_2Ph)_4$ by neutron diffraction. (Ref. 60)

Pentuply H-bridged metal-metal bonds, $M(\mu\text{-H})_5M$, are unknown, but their existence has been speculated upon [189]. It was concluded that steric factors are probably more important than electronic factors in determining the stability of such compounds. In any $M(\mu\text{-H})_nM$ system, it is the set of non-bonding $H \cdots H$ contacts (generally around 1.9 Å) which determines the size of the $(H)_n$ ring, which necessarily gets larger along the series $M(\mu\text{-H})_3M$, $M(\mu\text{-H})_4M$, $M(\mu\text{-H})_5M$. Thus, the possible existence of $M(\mu\text{-H})_5M$-bonded molecules may be limited to metals having very large covalent radii, large enough to overlap efficiently with the H_5 ring.

IV. Mixed Bridged Systems [e.g., $M(\mu\text{-H})(\mu\text{-}OR)M$, $M(\mu\text{-H})(\mu\text{-}X)_2M$, etc.]

Mason [190] and Churchill [5,191–194] have made extensive investigations of the systematic changes that take place when μ-H ligands are replaced by other bridging groups (e.g., halide, alkoxide or thiolate ligands). In the original paper by Mason [190], an increase in M−M distance was noted along the series $Os(\mu\text{-H})_2Os$, $Os(\mu\text{-H})(\mu\text{-SEt})Os$, $Os(\mu\text{-OMe})_2Os$, which was attributed to a decrease in formal bond order (two, one, zero). This work was later extended by Churchill and co-workers to a large number of rhodium [5,191] iridium [5,192] and osmium [193,194] complexes, which showed essentially the same trends.

Changes of this sort have also been noticed in complexes having three bridging ligands. One sees a shortening, for example, in metal-metal distance in going from $[X_3Mo(\mu\text{-X})_3MoX_3]^{3-}$ $(X = Cl, Br)$ to $[X_3Mo(\mu\text{-X})_2(\mu\text{-H})MoX_3]^{3-}$ [195]. The chloro species, $[Cl_3Mo(\mu\text{-Cl})_2(\mu\text{-H})MoCl_3]^{3-}$, incidentally, has been accurately analyzed by X-ray methods on two different salts [196,197], one of which is a highly unusual mixed salt $[Et_4N]_3^+[H_5O_2]^+[Mo_2Cl_8H]^{3-}[MoCl_4O(H_2O)]^-$ [197].

E. Metal Clusters

Originally considered chemical curiosities, polynuclear transition metal complexes are receiving considerable attention these days [198–201]. The aggregation of transition metal fragments is now recognized as a facile process under the proper conditions. Interest in these compounds stems, in part, from the possibility that clusters might serve as convenient homogeneous catalysts, and the hope that they might act as models of metal surfaces. Additionally, of course, these compounds are interesting in their own right: the fascinating geometrical forms that metal clusters exhibit [198] are often unknown in other branches of chemistry.

Although there have been numerous crystal structure determinations on cluster complexes, the problems of locating a hydrogen atom on a cluster surface with X-rays are understandably more severe than in the case of mononuclear complexes. Since a hydrogen ligand in a cluster can bridge two, three, or more metal atoms, it is situated in what might be described as a "fog" of electron density. As a consequence of this difficulty, structure determinations of cluster hydrides have, in the past, relied heavily upon indirect methods of locating the hydrogen ligands. It is perhaps for these compounds, more so than with other metal hydrides, that neutron diffraction offers the most obvious advantage.

I. Edge-bridging (μ_2-H) Systems

The edge-bridging position, based on structural work carried out to date, is undoubtedly the most common site of attachment of a hydrogen atom to a metal cluster. Earlier in this article we described X-ray work carried out on $H_3Mn_3(CO)_{12}$ (Figs. 3 and 4).[18]. A very recent neutron diffraction study on the related rhenium species $H_3Re_3(CO)_8[(EtO)_2POP(OEt)_2]_2$ [202] confirms the planar, equatorial positioning of the three H atoms. The average molecular parameters found for this compound [Re–H = 1.81(1) Å, Re–H–Re = 130(1)°] compare very well with the X-ray derived values for $H_3Mn_3(CO)_{12}$ [Mn–H = 1.72(3) Å, Mn–H–Mn = 131(7)°].

Neutron diffraction studies on $H_3Rh_3[P(OMe)_3]_6$ [182], however, show a very different geometry (Fig. 47). In this case the distribution of bridging H atoms is skewed: one is in the plane of the Rh_3 ring, one is above, and the other is below the Rh_3 plane (each by about 1 Å). The arrangement of the individual H_2RhP_2 units (which are approximately square planar) relative to the equatorial Rh_3 plane is also skewed: two of them are tilted 35° and 38° in opposite directions with respect to the Rh_3 ring, while the third makes an angle of 64° to it. The structural difference between $H_3Mn_3(CO)_{12}$ and $H_3Rh_3[P(OMe)_3]_6$ could be related to their different electron configurations: the former is a saturated 48-electron cluster, while the latter is unsaturated (42 electrons).

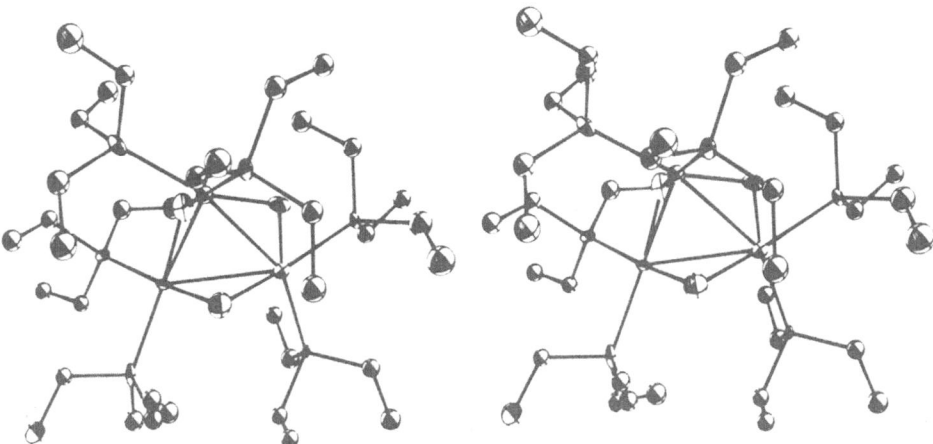

Fig. 47. A stereoscopic diagram of the structure of $H_3Rh_3[P(OMe)_3]_6$, as determined by neutron diffraction (Ref. 182a). Note that, in contrast to $H_3Mn_3(CO)_{12}$ (Fig. 3), the H atoms in this molecule are not all coplanar with the M_3 plane: one is in plane, another is above, and the third is below

$H_2Os_3(CO)_{10}$, one of the few unsubstituted hydridocarbonyl clusters that easily forms large crystals, has the distinction of having been analyzed four times[d] by diffraction methods: twice by X-rays [203,204] and twice by neutrons [30,184]. Its structure, which has been described earlier in this article (Fig. 44, Sect. D. II.), is an example of a metal cluster complex containing a doubly H-bridged edge [i.e., $M(\mu\text{-}H)_2M$]. Other complexes believed to contain this feature are $[H_3Re_3(CO)_{10}]^{2-}$ [185], $[H_4Re_3(CO)_{10}]^-$ [186], $H_2Os_3(CO)_9(CNBu^t)$ [205] and $H_2Os_3(CO)_9(PPh_3)$ [206].

Triosmium carbonyl clusters have the advantage of forming stable crystalline complexes with a wide variety of organic molecules (or fragments of molecules); consequently, the literature on their structures is voluminous (see Table 5). Among hydride complexes, those that have been analyzed by neutron diffraction include $HOs_3(CO)_{10}(C_2H_3)$ [30], $H_2Os_3(CO)_9S$ [207], $H_2Os_3(CO)_{10}(CH_2)$ [208] and $HDOs_3(CO)_{10}(CHD)$ [25]. The structure determination of $HOs_3(CO)_{10}(C_2H_3)$ [30] represents an application of combined X-ray/neutron refinement (Sect. A. III.). Its structure shows a vinyl group and a H atom bridging the same side of the Os_3 triangle, with the vinyl group σ-bonded to one Os atom and π-bonded to the other. In the structure of $H_2Os_3(CO)_{10}(CH_2)$ [208] (Fig. 48), one Os—Os edge is doubly bridged $[Os(\mu\text{-}H)(\mu\text{-}CH_2)Os]$, one is singly-bridged $[Os(\mu\text{-}H)Os]$, while the third is unbridged. This structure represents the first accurate determination of a coordinated methylene ligand [H—C—H = 106.0(8)°]. The structural analysis of the partially deuterated analogue, $HDOs_3(CO)_{10}(CHD)$ [25], shows that the H atoms have a distinct preference for the metal-bound positions, while the D atoms prefer the methylene positions. This distribution can be rationalized in terms of the difference in zero-point energies between the C—H (high energy) and Os—H (low energy) stretching vibrations.

Other noteworthy trinuclear hydride complexes include $HMn_3(CO)_{10}(BH_3)_2$ [121], a rare cluster containing both M—H—M and M—H—B bridges (Fig. 25), and the com-

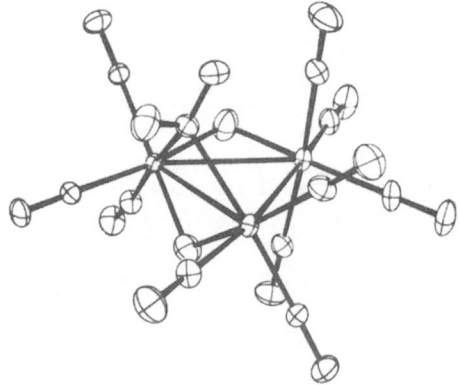

Fig. 48. The structure of $H_2Os_3(CO)_{10}(CH_2)$, again determined by neutron diffraction (Ref. 208). One Os—Os bond is doubly bridged $[Os(\mu\text{-}CH_2)(\mu\text{-}H)Os]$, one is singly bridged $[Os(\mu\text{-}H)Os]$, and the third is unbridged

---

[d] Actually, it was almost analyzed five times. The authors were about to begin neutron data collection on a particularly fine specimen of $H_2Os_3(CO)_{10}$ when word came that two other groups were working on the problem [30,184]

plexes $H_2Os_3(CO)_{11}$ [78] (Fig. 14), and $H_2Os_3(CO)_{11}(PPh_3)$ [209], which are uncommon examples of metal clusters containing terminal H ligands.

$H_4Ru_4(CO)_{12}$ and its derivatives constitute an interesting case of isomerization among various edge-bridged structures. The complexes show a characteristic four-long/two-short pattern of M–M distances, corresponding to M–H–M and M–M bonds respectively, but the arrangement of these bonds can vary from compound to compound. For example, $H_4Ru_4(CO)_{12}$ [210], $H_4Ru_4(CO)_{11}[P(OMe)_3]$ [210] and $H_4Ru_4(CO)_{10}(PPh_3)_2$ [210,211] all show the two short bonds opposite each other (*XXIII*), while in $H_4Ru_4(CO)_{10}(Ph_2PCH_2CH_2PPh_2)$ [212] they are adjacent to each other (*XXIV*). In what is perhaps an even more spectacular example, two different isomers of $[(Ph_3P)_2N]^+[H_3Ru_4(CO)_{12}]^-$ [213] were isolated under apparently identical conditions. One isomer has a pattern of bonds arranged in a C_{3v} manner (*XXV*), while the other has C_2 symmetry (*XXVI*). The existence of this rich variety of geometrical forms *XXIII–XXVI* clearly indicates that the energy differences among the various structures are small. Indeed, NMR studies in solution by Kaesz [214], Shapley [212a] and co-workers had earlier shown that H exchange between the various edge-bridging sites is a facile process.

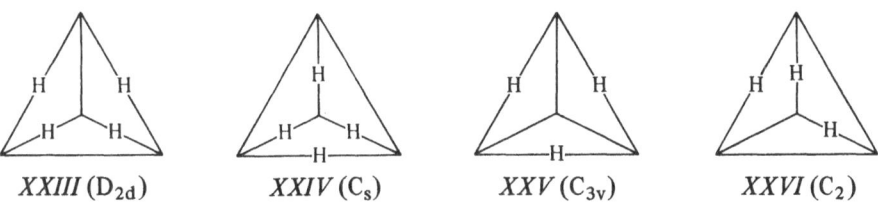

XXIII (D_{2d})   *XXIV* (C_s)   *XXV* (C_{3v})   *XXVI* (C_2)

The D_{2d} form (*XXIII*) has subsequently been shown to exist in two other structures. It has been found in $H_4Ru_4(CO)_8[P(OMe)_3]_4$ by a single-crystal neutron diffraction analysis [202], which yielded average Ru–H and Ru–H–Ru parameters of 1.773(2) Å and 114.2(3)° respectively. It also exists in the unsaturated, 58-electron complex $[H_4Rh_4(C_5Me_5)_4]^{2+}$ [215] which, in contrast to the saturated 60-electron $H_4Ru_4(CO)_{12}$ family, contains static rather than fluxional H atoms. Finally, the C_s structure (*XXIV*) has been found in $[H_4Re_4(CO)_{13}]^{2-}$ [36].

Other tetranuclear edge-bridged hydride complexes which have been structurally characterized include $H_2Ru_4(CO)_{13}$ [216] and $H_2FeRu_3(CO)_{12}$ [217], which show a two-long/four-short pattern of M–M distances (*XXVII*), and $[H_6Re_4(CO)_{12}]^{2-}$ [14,218], in which all six M–M edges are H-bridged (*XXVIII*). Several papers, incidentally, have commented on the fact that the M–H–M planes can take various orientations relative to the M–M–M planes [219–221].

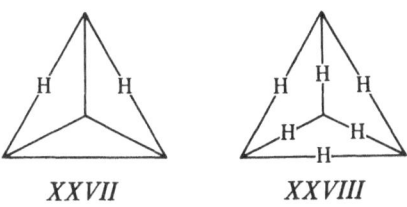

XXVII   *XXVIII*

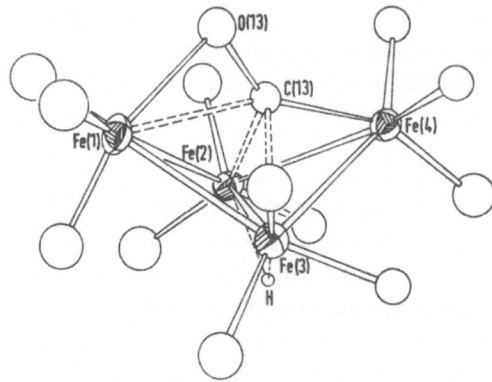

Fig. 49. An illustration of the unusual cluster [HFe₄(CO)₁₃]⁻, showing the unique four-electron carbonyl ligand, which binds in both a σ- and π-fashion (Ref. 152). The oxygen atoms of the CO groups have been removed for clarity

[HFe$_4$(CO)$_{13}$]$^-$ [152] (Fig. 49) is an unusual example of an electron-excess (62 e$^-$) cluster. In this molecule a unique carbonyl group is found which serves both as a σ- and a π-donor, and hence can be considered a ligand which supplies four electrons to the cluster. The carbon atom of this CO group is bonded to all four iron atoms, while its C–O axis is oriented in a π-fashion (i.e., sideways) with respect to Fe(1). The H atom is known [152b] to bridge the Fe–Fe bond opposite this quadruply-bridging CO group, but the presence of two ^{1}H NMR signals suggests the existence of isomers in solution.

Interestingly, the isoelectronic iron/ruthenium analogue [HFeRu$_3$(CO)$_{13}$]$^-$ [222] has a different structure: a tetrahedral geometry with normal CO groups. In a recent neutron diffraction study, the H atom was located on one of the Ru–Ru edges, making Ru–H and Ru–H–Ru parameters of 1.821(3) Å and 106.4(2)°, respectively.

In the terminology used in Wade's rules [199], the 60-electron tetrahedral cluster [HFeRu$_3$(CO)$_{13}$]$^-$ is considered a *closo* species (*XXIX*) and the 62-electron "open butterfly" structure of [HFe$_4$(CO)$_{13}$]$^-$ is considered a *nido* species (*XXX*). The structural difference is related to the fact that one of the CO groups is a 2-electron donor in the former and a 4-electron donor in the latter.

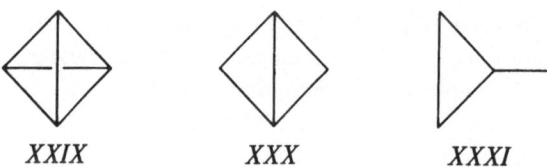

XXIX   *XXX*   *XXXI*

The process *XXIX* to *XXX* formally corresponds to the breaking of a metal-metal bond. If this is carried one step further, the 64-electron structure *XXXI* would result, one that might be considered *arachno*. One example of structure *XXXI* is illustrated by [H$_4$Re$_4$(CO)$_{15}$]$^{2-}$ [79] (Fig. 50). This unusual molecule contains three different types of H atoms: one type that bridges the edges of the Re$_3$

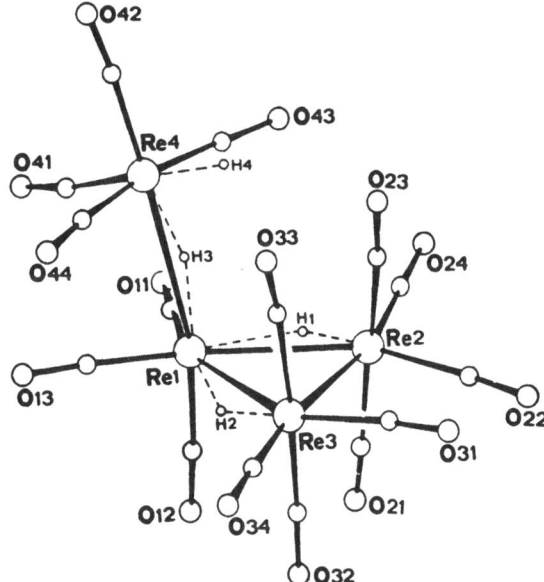

Fig. 50. A schematic diagram of $[H_4Re_4(CO)_{15}]^{2-}$, with the suspected hydrogen positions. Note that there are three different kinds of hydride ligands in this molecule (Ref. 79)

triangle, one that bridges the bond from Re_3 to the exocyclic Re atom, and finally a third that is terminally bonded to the exocyclic Re atom.

The cyclohexenyl complex $H_3Os_4(CO)_{11}(C_6H_9)$ [223] represents a molecule in which the predicted H positions do not all correspond to long metal-metal distances. The six edges of the tetrahedron have lengths of 2.984(1), 2.953(1), 2.937(2), 2.817(1), 2.794(1) and 2.793(1) Å. Using potential energy calculations based on a consideration of non-bonding contacts, Orpen [37] predicted that H atoms would bridge the first, second and fourth edges of the above list. This prediction was indirectly confirmed by the neutron diffraction analysis of a related molecule, $H_3Os_4(CO)_{11}(CH=CHC_6H_5)$ [202], which showed H bridges on two long Os–Os edges [2.977(3), 2.950(2) Å] and one short Os–Os edge [2.787(3) Å].

A number of interesting pentanuclear hydride complexes have been reported by Lewis, Johnson, Sheldrick and co-workers. $[HOs_5(CO)_{15}]^-$ [224] has a trigonal bipyramidal arrangement of osmium atoms, while $H_2Os_5(CO)_{16}$ has a novel pentanuclear skeleton based on an edge-bridged tetrahedron [225]. Both clusters are believed to contain M–H–M bridges. The pyrolysis of $Os_3(CO)_{11}[P(OMe)_3]$ has yielded two interesting carbide/hydride clusters, $HOs_5C(CO)_{13}[OP(OMe)OP-(OMe)_2]$ [226a] and $HOs_5C(CO)_{14}[OP(OMe)_2]$ [226b]. They have geometries that can be described as intermediate between square pyramidal and trigonal bipyramidal, with a five-coordinate carbon atom at the center (Fig. 51). A corresponding interstitial nitride complex $HFe_5N(CO)_{14}$ has recently been discovered [227].

An intriguing cluster compound synthesized a decade ago is $H_6Cu_6(PPh_3)_6$ [228]. It has been investigated via X-ray diffraction and the results are illustrated in Fig. 52. The geometry of the molecule shows a pattern of six long and six short Cu–Cu bonds, which implies the presence of six edge-bridging hydrogen atoms on two opposite faces of the octahedron. This molecule is the only hydrido copper cluster

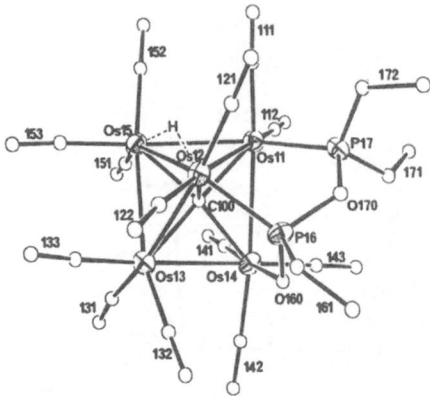

Fig. 51. The unusual hydride/carbide cluster $HOs_5C(CO)_{13}[OP(OMe)OP(OMe)_2]$, formed by the pyrolysis of $Os_3(CO)_{11}[P(OMe)_3]$ (Ref. 226 a). The carbon atom is situated in a site which appears from this viewpoint to be square pyramidal, but is actually mid-way between square pyramidal and trigonal bipyramidal

reported to date. Compared with other hexanuclear clusters, which usually possess 86 electrons, this 84-electron cluster may be considered electron-deficient. One noteworthy feature of this compound is the fact that the H atoms are undetectable in the IR and proton NMR spectra of the compound. Their presence could be chemically verified only by decomposition with acid to yield the corresponding amount of H_2.

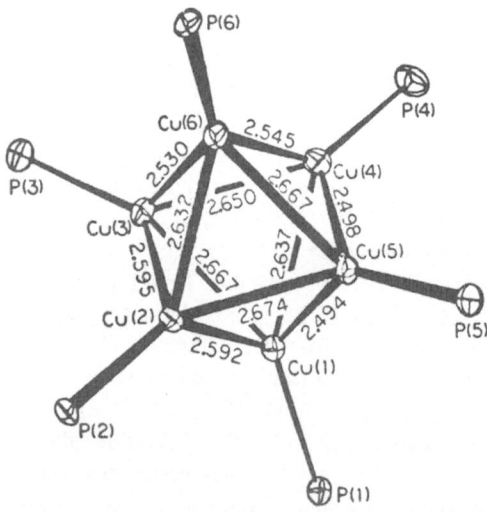

Fig. 52. A diagram of $H_6Cu_6(PPh_3)_6$ with the phenyl rings omitted for clarity. The hydrogen ligands are believed to bridge the six long Cu–Cu vectors, which define two opposite triangular faces of the cluster (Ref. 228)

II. Face-bridging (μ_3-H) Systems

The face-bridging (or three-coordinate) mode of hydrogen binding to a metal cluster appears to be much less common than the edge-bridging mode. At present, there are only about a dozen compounds known or suspected to contain such linkages.

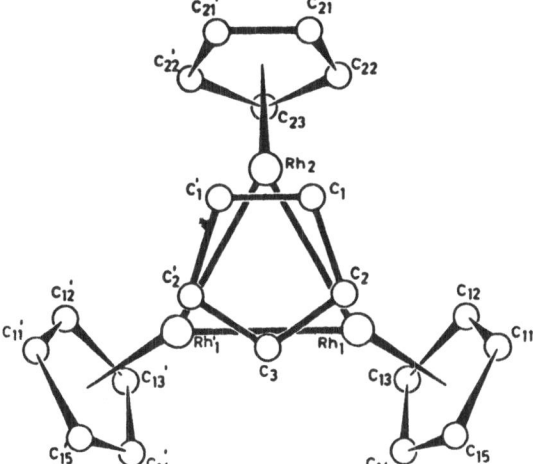

Fig. 53. The first example of a face-bridging H atom is believed to exist in HRh$_3$Cp$_4$. One of the four cyclopentadienyl rings in this compound is coordinated in a most unusual fashion, parallel to the Rh$_3$ plane, while the μ_3-H ligand (not located) is believed to bridge the other side of the Rh$_3$ triangle (Ref. 229)

The first report of a face-bridging H atom was published in 1968 by Mills and Paulus[229], who described the structure of HRh$_3$(C$_5$H$_5$)$_4$ (Fig. 53). The molecule consists of a triangle of rhodium atoms, each of which is attached to a cyclopentadienyl ring. This triangle is bridged on one side by a most unusual μ_3-C$_5$H$_5$ ring, and on the other side, it is believed, by the μ_3-H atom. Other examples of trinuclear complexes with face-bridging H atoms are [H$_7$Ir$_3$(PCy$_3$)$_3$(py)$_3$]$^{2+}$ [80] (6 σ-H + 1 μ_3-H), and {H$_7$Ir$_3$[Ph$_2$P(CH$_2$)$_3$PPh$_2$]$_3$}$^{2+}$ [81] (3 σ-H + 3 μ_2-H + 1 μ_3-H). In none of these compounds were the H atoms actually located.

The unsaturated (56 e$^-$) molecule H$_4$Re$_4$(CO)$_{12}$ [12] (Fig. 2) was discussed earlier in connection with methods of locating H atoms with X-ray data (Sect. A. II.). A symmetry-averaged H position was located from a composite difference Fourier map (Fig. 6), which corresponds to an unrefined Re–H distance of 1.77 Å. This distance is, as is the case with most M–H bond lengths derived from X-ray data, probably 0.1–0.2 Å shorter than its true value.

The saturated (60 e$^-$) tetrahedral cluster H$_4$Co$_4$(C$_5$H$_5$)$_4$ and the electron-excess (63 e$^-$) tetrahedral cluster H$_3$Ni$_4$(C$_5$H$_5$)$_4$ were analyzed with X-rays by Huttner and Lorenz[230,231]. In the former case the face-bridging H atoms were located and were found to be displaced an average of 0.8 Å from the Co$_3$ faces of the tetrahedron (Fig. 54). The average measured Co–H distance was 1.67(7) Å. In the X-ray analysis of H$_3$Ni$_4$(C$_5$H$_5$)$_4$ the H atoms were not found, but are also believed to be face-bridging, partly in analogy with the cobalt cluster and partly on the basis of molecular distortions (three of the cyclopentadienyl rings seemed to be tilted away from the fourth). This prediction was subsequently confirmed by a neutron diffraction analysis[232], which revealed a H$_3$Ni$_4$ core that resembles a cube with a missing corner (Fig. 55). H$_3$Ni$_4$(C$_5$H$_5$)$_4$, incidentally, is a rare example of a paramagnetic organometallic cluster (it has three unpaired electrons).

A theoretical treatment of tetranuclear metal hydride clusters has been published by Hoffmann and co-workers[233]. In that article, clusters with various electron configurations (56, 60, 63 electrons) are discussed, and a rationalization is given for

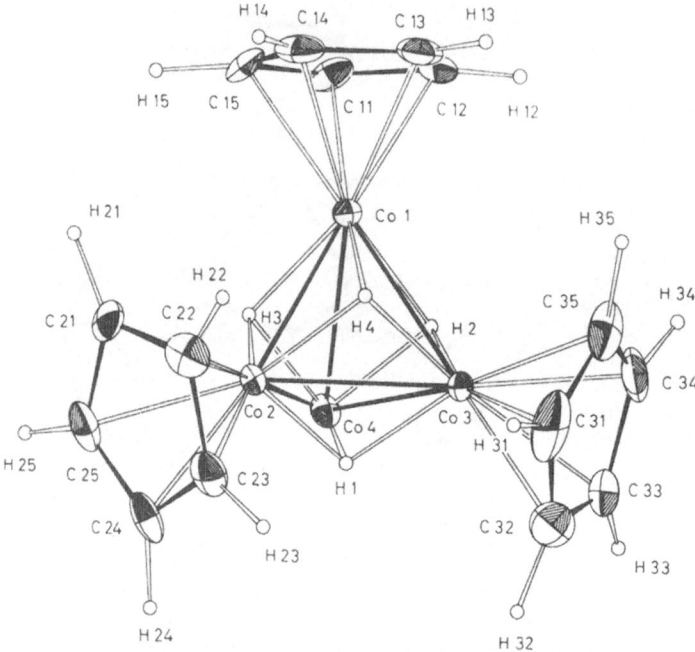

Fig. 54. The structure of H$_4$Co$_4$Cp$_4$, indicating the face-bridging hydrogen positions as determined from an X-ray study (Ref. 230)

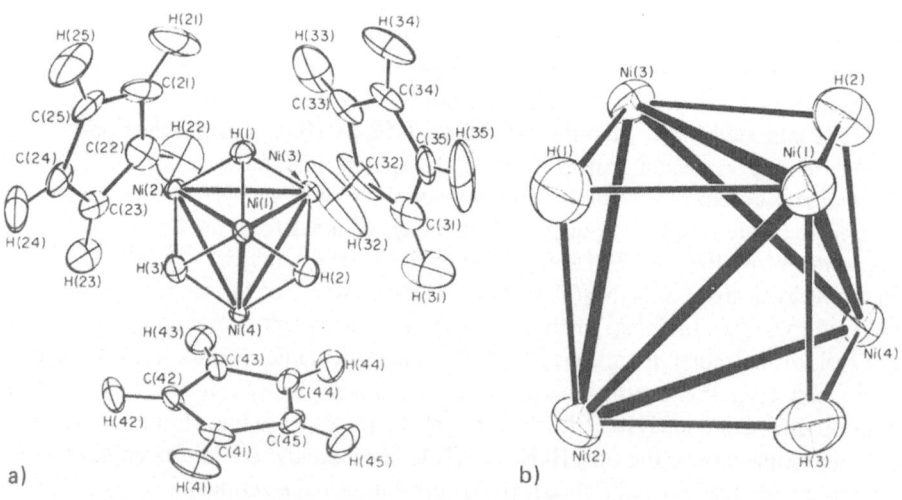

a)   b)

Fig. 55. a The structure of H$_3$Ni$_4$Cp$_4$, determined by a neutron diffraction study (Ref. 232). One of the C$_5$H$_5$ rings has been removed for clarity. **b** The H$_3$Ni$_4$ core of this molecule, which resembles a cube with a missing corner

the seemingly odd 63-electron configuration of $H_3Ni_4(C_5H_5)_4$. Also presented is an analysis of the energy differences between various carbonyl orientations [i.e., eclipsed (I) vs. staggered (II)], and between edge-bridging and face-bridging hydrogen arrangements. In an unrelated approach, Green, Mingos and Seddon[42] have developed a set of rules, modeled after the "styx" rules of the boron hydrides[43], that were used to predict the number of edge-bridging and face-bridging H atoms in several metal cluster complexes.

The history associated with the mixed metal cluster $HFeCo_3(CO)_{12}$ and its derivatives is rather interesting. First synthesized by Chini in 1960[234], the molecule had been the subject of several spectroscopic investigations aimed at determining the location of the hydride ligand. Inelastic neutron scattering experiments[235] and mass spectral studies[236] suggested the H atom to be inside the metal tetrahedron, but this model was disproved in an X-ray study by Huie, Knobler and Kaesz[237], which showed the H atom to be outside the cluster, capping the Co_3 face. More accurate molecular parameters [Co–H = 1.734(4) Å, Co–H–Co = 91.8(2)°] were later obtained from a single-crystal neutron diffraction study[238] (Fig. 56). The isoelectronic $[HFe_3Ni(CO)_{12}]^-$ anion, interestingly enough, has the H atom capping one of the Fe_2Ni faces rather than the Fe_3 face[239].

Two octahedral clusters that are believed to contain face-bridging H atoms are $H_2Ru_6(CO)_{18}$[240] and $[HOs_6(CO)_{18}]^-$[241]. In each case, the H atoms were not directly located but their positions were revealed by the presence of enlarged faces. In $H_2Ru_6(CO)_{18}$ (Fig. 57), for example, the H-capped triangles, situated at opposite sides of the octahedron, have Ru–Ru distances in the range 2.950–2.959 Å, while the other Ru–Ru distances in the cluster lie in the range 2.858–2.874 Å. (Interestingly, the isoelectronic $H_2Os_6(CO)_{18}$ has a completely different structure from $H_2Ru_6(CO)_{18}$, as will be seen in the next section.) The structure of $H_2Ru_6(CO)_{18}$ provides another good example of how the orientation of carbonyl groups can often give useful clues about the positions of the H atoms. A close examination of Fig. 57

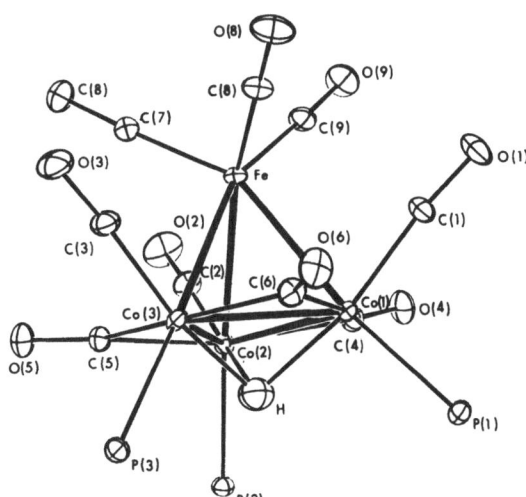

Fig. 56. The molecular structure of $HFeCo_3(CO)_9(P(OMe)_3)_3$, with methoxy groups removed for clarity (Ref. 238). The hydride ligand is clearly shown symmetrically bridging the Co_3 face, and is displaced 0.978 (3) Å from it

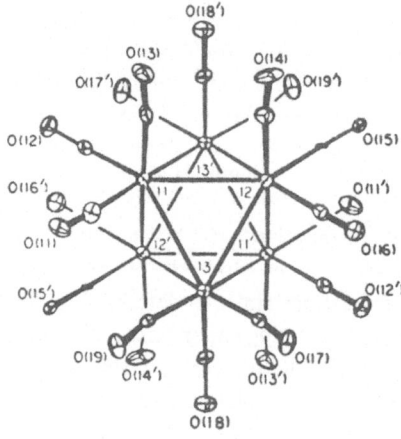

Fig. 57. The molecular geometry of $H_2Ru_6(CO)_{18}$, with two enlarged faces [labeled (11, 12, 13) and (11', 12', 13')] signifying the presence of triply-bridging H atoms (Ref. 240). The orientation of the CO groups also reveals the positions of the "invisible" μ_3-H atoms (see text)

reveals that the CO groups fall neatly into two classes. Some of them are approximately collinear with the Ru-Ru bonds (i.e., the CO groups that are labeled 11, 13, 14, 16, 17, 19), while the others (labeled 12, 15, 18) are not. Those of the latter set, in fact, are oriented in such a way that they "point" to the presumed face-capping H atoms situated above the Ru (11, 12, 13) triangular face.

Several examples of H atoms capping M_2B triangular faces have been discovered by Grimes and co-workers. In the X-ray study of $(C_5H_5)_2Co_2(B_4H_6)$ (Fig. 58), the H atoms were found to be displaced 0.78 Å from the Co_2B faces, making Co-H and B-H distances of 1.55 and 1.51 Å respectively [114]. Other metalloborane complexes in which μ_3-H atoms are believed to exist include $Co_3(B_3H_5)(C_5H_5)_3$ [242], $(C_5H_5)_2Fe(H)Co(Me_2C_2B_3H_3)$ [243] and $H_2Fe(MeC_2B_4H_4)_2$ [244]. In the latter compound, the H atoms are thought to bridge FeB_2 faces [244].

The existence of triply-bridging H atoms (and other types of H bridges) is often encountered in systems involving chemisorbed hydrogen (e.g., Ref. 245). Thus, one rationale for studying the molecular dimensions of covalent metal hydride clusters is to provide reasonable estimates for corresponding distances and angles involving H atoms adsorbed on a metallic surface.

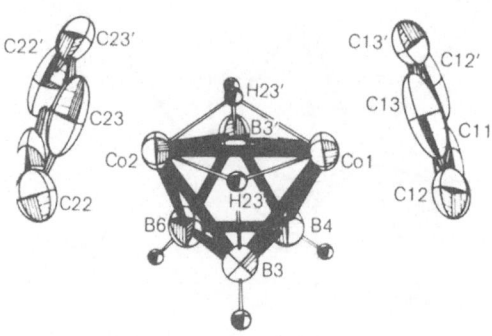

Fig. 58. An illustration of the structure of $Cp_2Co_2B_4H_6$, showing the first example of a triply-bridging H atom situated on a M_2B face (Ref. 114)

III. Systems with Interstitial Hydrogen Atoms

The existence of H atoms of a higher coordination number (four, five or six) is an intriguing problem. It was not until very recently that examples of such species were unequivocally established in molecular complexes [246,247]. In the solid state literature, however, high-coordination H atoms are well-known entities: there are many examples of binary metal hydrides in which H atoms are known to occupy tetrahedral or octahedral sites in a metal lattice [8].

The first report of an interstitial H atom in a non-metallic solid appeared as a publication by Simon in 1967, on the structure of polymeric HNb_6I_{11} [248]. In this classic study, the neutron powder patterns of HNb_6I_{11} and DNb_6I_{11} were compared, and the differences were interpreted to indicate the presence of six-coordinate H atoms in the center of the Nb_6 octahedra. This piece of work demonstrates that even powdered crystalline samples can yield useful molecular information if the structure in question is simple enough. Although a Nb–H distance in HNb_6I_{11} was not explicitly given, it can be calculated, based on the known Nb–Nb distance of 2.84 Å in the Nb_6 cluster, to be 2.01 Å.

The probable existence of an interstitial H atom in a molecular (i.e., non-polymeric) complex was first suggested in a paper by Eady, Johnson, Lewis and co-workers on $[HRu_6(CO)_{18}]^-$ [249]. In the X-ray study of this anion, the H atom was not directly located, but was presumed to be in the center of the metal cage on the basis of the near-perfect octahedral symmetry of the cluster. There were no discernible distortions of the cluster or its ligands which might have indicated an alternative placement of the H atom.

Unequivocal proof for the existence of a six-coordinate H atom in an octahedral cluster was subsequently provided by us, in collaboration with Dr. Thomas F. Koetzle of Brookhaven National Laboratory and the late Professor Paolo Chini of the University of Milan, in a single-crystal neutron diffraction study of $[HCo_6(CO)_{15}]^-$ [246] (Fig. 59). Ironically, this discovery was somewhat accidental, since the H atom in that cluster, which has an anomalously low NMR chemical shift ($\tau = -13.2$ ppm), was originally thought to be situated on the carbonyl ligands [250]. One other unusual feature of the interstitial H atom in this cluster is that it exchanges readily with proton-accepting solvents [246], in complete contrast to other interstitial atoms (like carbon atoms in metal carbide clusters), which are invariably held very tightly within the metal cage. The high mobility of the H atom in and out of the cluster is reminiscent of the rapid migration of H atoms within metal lattices [8]. The H atom in $[HCo_6(CO)_{15}]^-$ was found to be, within experimental error, at the geometric center of the cluster, making an average Co–H distance [1.824(13) Å] which is distinctly longer than that in $HFeCo_3(CO)_9[P(OMe)_3]_3$ [1.734(4) Å] [238].

Jackson, Johnson, Lewis and co-workers very recently carried out a single-crystal neutron diffraction study on $[AsPh_4]^+[HRu_6(CO)_{18}]^-$ [251] and confirmed their original suggestion of an interstitial H atom. This complex also has an anomalously low ^1H NMR chemical shift ($\tau = -6.4$ ppm) [249].

Interstitial H atoms have also been found in larger metal clusters. In a neutron diffraction study of $[HNi_{12}(CO)_{21}]^{3-}$ and $[H_2Ni_{12}(CO)_{21}]^{2-}$, Williams, Dahl, Chini and co-workers found H atoms lodged in octahedral cavities of these multi-hole

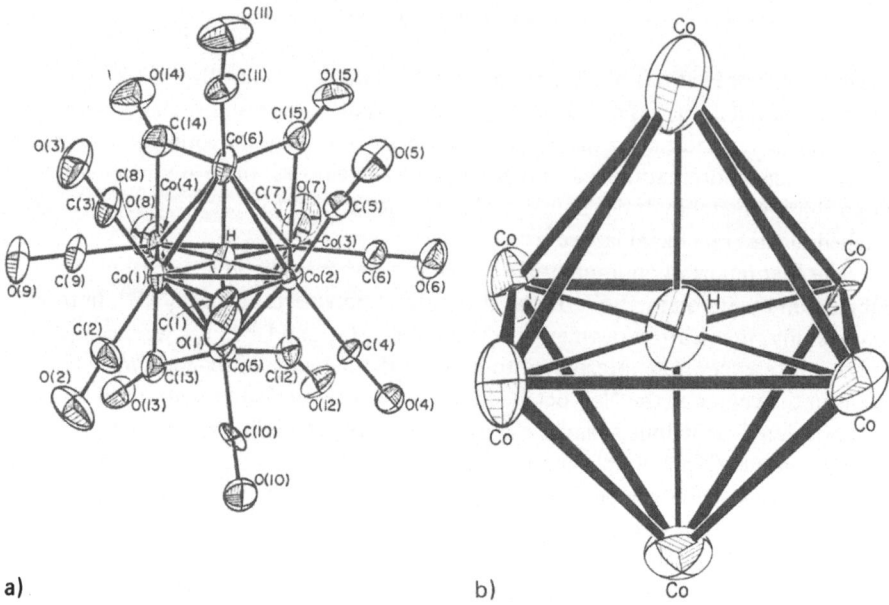

a)   b)

Fig. 59. a An ORTEP plot of $[HCo_6(CO)_{15}]^-$, in which a six-coordinate H atom was discovered using single-crystal neutron diffraction techniques (Ref. 246). **b** A close-up view of the core of $[HCo_6(CO)_{15}]^-$, showing the near-perfect centering of the interstitial H atom within the octahedral cavity

clusters[247] (Fig. 60). In contrast to the HM_6 clusters, however, the H atoms are asymmetrically positioned: they are significantly displaced towards one of the interior triangular faces of the octahedral cavities (Fig. 61). This indicates that there is actually more than enough room in an octahedral hole to accommodate an H atom. For these clusters, the ^1H NMR chemical shifts are in the "normal" range (34 τ for $[HNi_{12}(CO)_{21}]^{3-}$ and 28 τ for $[H_2Ni_{12}(CO)_{21}]^{2-}$), which suggests that the chemical shifts of interstitial H atoms may be highly sensitive to the exact placement of the H atoms within the cavity: i.e., whether they are situated on a site of octahedral symmetry or not.

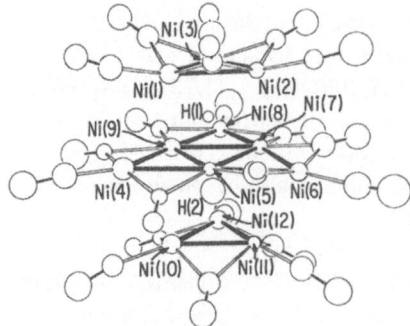

Fig. 60. The structure of the $[H_2Ni_{12}(CO)_{21}]^{2-}$ ion, showing the two interstitial H atoms in octahedral cavities. Note that the H atoms are slightly displaced towards the central Ni(5,7,9) triangle of the cluster (Ref. 247)

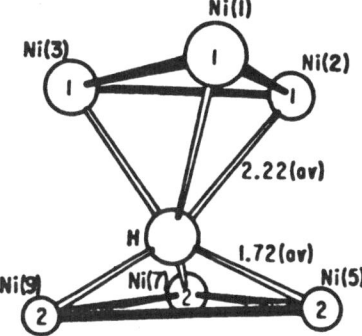

Fig. 61. A blow-up of the filled octahedral cavity of the $[HNi_{12}(CO)_{21}]^{3-}$ ion, showing the highly distorted positioning of the interstitial H atom (Ref. 247). This H atom appears to be approaching a triply-bridging condition, capping a triangular face of an octahedron from within

Examples of four- or five-coordinate H atoms are much less well-defined. As mentioned earlier, H atoms in tetrahedral cavities are often found in the structures of binary metal hydrides[8], but there are as yet no indications of their existence in discrete molecular complexes. An interstitial H atom was suspected to exist in $HFeCo_3(CO)_{12}$ [235,236], but in retrospect (now that we know the dimensions of this metal cage) it is clear that there simply is not enough room inside the $FeCo_3$ tetrahedron to accommodate an H atom[238]. A tetrahedral cavity is inherently much smaller than an octahedral one, for a given metal-metal distance. Indeed, we speculate that only metals with very large covalent radii could form tetrahedral cavities large enough to accommodate an H atom.

The other possible binding mode for a four-coordinate H atom is a quadruply-bridging one; i.e., one which caps a square face of metal atoms. For chemisorbed H atoms on certain metal surfaces this geometric arrangement appears to be stable, on the basis of experimental[252] and theoretical[253] investigations. For molecular complexes, however, such quadruply-bridging H atoms would be harder to find, since most clusters form polyhedra with triangular rather than square faces. In the structure of $H_2Os_6(CO)_{18}$ [241] (Fig. 62), one of the few examples of a non-octahedral hexameric cluster, it was originally suspected that one of the two H atoms might be bridging the basal square face. But subsequent potential energy calculations[37] favor edge-bridging positions for both H atoms. There is a remote possibility that square-capping H atoms might be found in the larger Rh_{13} clusters which will be discussed next.

For five-coordinate hydrogen, one again has two choices: it can on principle exist within a trigonal bipyramidal or a square pyramidal cavity. Of the two, the latter is the more likely possibility since its hole size is larger (a square pyramidal cavity is roughly on octahedral hole with one vertex missing). At present, the most likely hosts for five-coordinate H atoms are the large Rh_{13} clusters, $[H_2Rh_{13}(CO)_{24}]^{3-}$ and $[H_3Rh_{13}(CO)_{24}]^{2-}$, synthesized a few years ago by Martinengo, Chini and co-workers. The cores of these clusters (Fig. 63) contain eight tetrahedral and six square pyramidal cavities. Not only are the latter favored as the H-containing sites on the basis of larger cavity size, but there is reasonably convincing experimental evidence that supports this point of view. Three independent X-ray structure determinations on $[(Ph_3P)_2N]_2[H_3Rh_{13}(CO)_{24}]$ [254], $[PPh_3Bz]_3[H_2Rh_{13}(CO)_{24}]$ [255], and

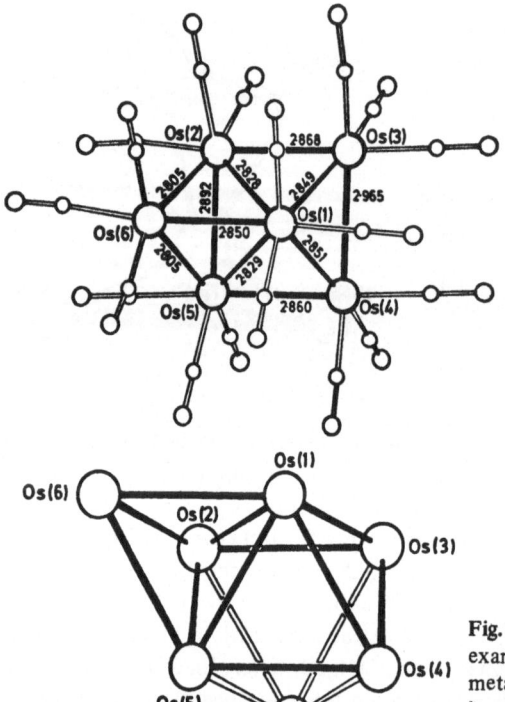

Fig. 62. Two views of $H_2Os_6(CO)_{18}$, a rare example of a non-octahedral hexanuclear metal cluster (Ref. 241). There was originally some suspicion that one of the H atoms might cap the basal square face of this cluster

$[NBu_4^n]_3[H_2Rh_{13}(CO)_{24}]$[256], show that there is a direct link between the number of H atoms and the number of expanded square pyramidal cavities. In $[H_2Rh_{13}(CO)_{24}]^{3-}$, for example, it has been found that two of the square pyramidal holes (average Rh–Rh distance 2.824 Å) are significantly larger than the other four (average Rh–Rh = 2.779 Å), while in $[H_3Rh_{13}(CO)_{24}]^{2-}$ there are three large (Rh–Rh = 2.83 Å) and three small (Rh–Rh = 2.78 Å) square pyramidal cavities[255]. Additionally, the newly-synthesized $[HRh_{13}(CO)_{24}]^{4-}$ anion also has a structure that fits this pattern: it has one large and five small holes[257].

Whether the H atoms in these square pyramidal cavities are considered four- or five-coordinate would depend on their location: if they are found to be entirely

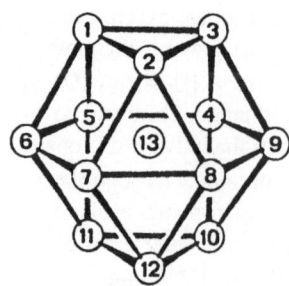

Fig. 63. The Rh_{13} skeleton of the clusters $[H_2Rh_{13}(CO)_{24}]^{3-}$ and $[H_3Rh_{13}(CO)_{24}]^{2-}$ (Ref. 254, 255). The H atoms could conceivably be located in the tetrahedral or square pyramidal cavities of this cluster, but X-ray evidence seems to indicate the latter (5-coordinate) sites as more likely (Ref. 255)

contained within the cavity (*XXXII*) or coplanar with the basal square plane of metal atoms (*XXXIII*) they are best considered five-coordinate, but if they turn out to be bridging the basal face from the outside (*XXXIV*) they might be considered four-coordinate. In the case of the Rh_{13} clusters, structure *XXXIV* would be unlikely because it would involve unreasonably large Rh–H distances.

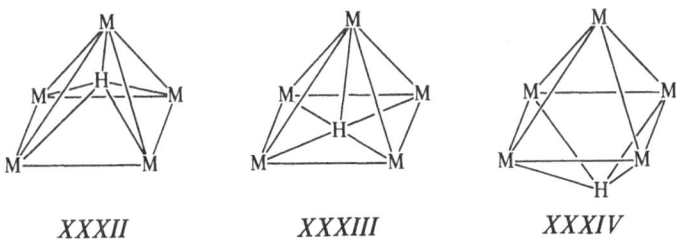

      XXXII   *XXXIII*   *XXXIV*

     Finally, eight-coordinate H atoms are theoretically possible since a cubic cavity is inherently much larger than an octahedral one. Although several cubic metal clusters are known [258], none of them is, as far as we know, suspected to contain interstitial H atoms.

F. Summary

In 1970, Frenz and Ibers reported the results of four neutron diffraction experiments in their structural review [1]. Since that time, over four dozen metal hydride complexes have been studied by this technique (see following tables). Based on these and supplemental X-ray results, some generalizations can be tentatively formulated:

(1) Terminal M–H distances involving first-row transition metals generally lie in the range of 1.4–1.7 Å; and for second and third row transition metals, in the range 1.5–1.8 Å.

(2) Bridging M–H distances (in M–H–M bonds) involving first-row transition elements usually lie in the range 1.6–1.9 Å; and for second and third row transition metals, in the range 1.7–2.1 Å.

(3) In cases where accurate numbers are available for both terminal and bridging M–H distances, one notices that the difference between these distances is rather constant: e.g., 0.16 Å for Mo, 0.15 Å for W, 0.20 Å for Re, and 0.17 Å for Os (see Table I). This suggests that a fairly reliable "adjustment factor" of 0.15–0.20 Å can be applied to estimate the value of a terminal M–H distance if the bridging M–H distance is known, or vice-versa. This value agrees rather well with the difference of 0.14 Å that one finds between terminal and bridging B–H distances [27,43].

(4) A noticeable decrease in M–H distances occurs as we move from left to right across a row of the periodic table. For example, terminal M–H distances show the following variation: Ta–H, 1.77(1) Å; W–H, 1.73(1) Å; Re–H, 1.67(1) Å, Os–H, 1.66(1) Å. This parallels the trend in covalent radii of these elements.

(5) There is insufficient information to make detailed comparisons between M—H—M and M_3H linkages, but we can estimate that triply-bridging M—H distances are probably equal to, or slightly longer than, bridging M—H distances.

(6) M—M distances in mono-hydrogen-bridged M—H—M bonds are about 0.10—0.45 Å longer than unbridged M—M bonds, provided that no additional bridging groups are present. The normal trend in M—M distances is: $M(\mu\text{-}H)M > M\text{—}M > M(\mu\text{-}H)_2M > M(\mu\text{-}H)_3M > M(\mu\text{-}H)_4M$. For a detailed discussion of this point, see reference 16.

(7) The shortest non-bonding H ... H contacts between adjacent hydride ligands appear to be about 1.85 Å.

Acknowledgements. We are very grateful to all our colleagues who took the trouble to read this manuscript and suggest improvements, and to those who generously supplied us with information prior to publication. This work was supported in part by the National Science Foundation and the Petroleum Research Fund (administered by the American Chemical Society).

Table 1. Average M–H Distances.[a] (For each element, the terminal distance is given on the first line, and the bridging distance on the second line)

Sc	Ti	V	Cr	Mn	Fe	Co	Ni	Cu
...	...	1.55(9)	...	1.59(2)b	1.54(3)	1.52(5)	1.45(9)	...
2.09(5)	1.85(8)	...	1.73(1)b	1.72(4)	1.70(1)b	1.72(1)b	1.69(1)b	1.71(1)b

Y	Zr	Nb	Mo	Tc	Ru	Rh	Pd	Ag
...	1.67(10)	1.69(4)	1.69(1)b	...	1.64(7)	1.57(8)	...	...
2.31(10)	2.21(4)	1.92(4)	1.85(1)b	...	1.79(1)b	1.79(2)b	...	2.19(5)

Hf	Ta	W	Re	Os	Ir	Pt	Au
...	1.77(1)b	1.73(1)b	1.67(1)b	1.66(1)b	1.67(7)	1.66(10)	...
2.10(1)b	...	1.88(1)b	1.87(1)b	1.83(1)b	1.78(5)b	1.78(10)	...

Th	U
2.03(1)b	...
2.29(3)b	2.36(2)b

[a] The weighted averages were computed using the equation $\bar{x} = \Sigma w_i x_i / \Sigma w_i$, where the weights are defined as $w_i = 1/\sigma_i^2$ and σ_i are the individual standard deviations. The standard deviation of the weighted average is given as $\bar{\sigma} = \sqrt{N/\Sigma w_i}$. For H positions measured off an X-ray difference Fourier map and not refined, an estimated standard deviation of 0.1 Å is assigned

[b] denotes a value derived in part from single-crystal neutron diffraction data

Table 2. A Tabulation of Individual M–H Distances

(A) *Terminal*

V	Cr	Mn	Fe	Co	Ni
1.55(9)		1.55(10)	1.49(6)	1.34(8)	1.37(10)
		[c]1.576(18)	1.51(4)	1.38(10)	1.48(7)
		[a]1.601(16)	[c]1.556(21)	1.38(16)	1.57(20)
		[a]$\overline{1.59(2)}$	1.57(12)	1.41(9)	1.45(9)
			$\overline{1.54(3)}$	1.43(6)	
				1.45(5)	
				1.53(15)	
				[c]1.556(18)	
				1.60(16)	
				1.65(12)	
				$\overline{1.52(5)}$	

Nb	Mo	Tc	Ru	Rh	Pd
1.62(10)	1.59(4)		1.36(10)		
1.69(4)	[a]1.685(3)		1.57(4)	1.36(8)	
1.70(3)	1.70(3)		1.58(7)	1.48(10)	
$\overline{1.69(4)}$	1.82(4)		1.60(8)	1.51(7)	
	[a]$\overline{1.69(1)}$		1.63(8)	1.54(9)	
			1.67(7)	1.60(12)	
			1.68(6)	1.65(8)	
			1.68(10)	1.66(5)	
			[b]1.69(4)	1.82(17)	
			1.70(15)	$\overline{1.57(8)}$	
			1.7(1)		
			1.7(1)		
			$\overline{1.8(1)}$		
			1.64(7)		

Ta	W	Re	Os	Ir	Pt
[a]1.774(3)	[a]1.728(7)	1.66(10)	1.52(7)	1.57(7)	1.66(10)
		[a]1.669(7)	1.64(6)	1.60(10)	
		[a]$\overline{1.68(1)}$	[a]1.659(3)	1.64(5)	
		[a]1.67(1)	[a]$\overline{1.66(1)}$	[b]1.70(4)	
				1.77(10)	
				1.77(12)	
				$\overline{1.82(17)}$	
				1.67(7)	

[a]   Value derived from single crystal neutron diffraction data
[b]   Value derived from powder neutron diffraction data
[c]   Value derived from gas phase electron diffraction data

Table 2 (continued)

(B) *Bridging*

Ti	V	Cr	Mn	Fe	Co	Ni	Cu
1.69(10)		[a]1.722(20)	1.65(10)	1.52(10)	1.40(4)	1.58(4)	[a]1.697(5)
1.75(8)		[a]1.734(9)	1.67(3)	1.56(6)	1.43(5)	[a]1.691(8)	1.82(3)
1.80(10)		1.78(6)	1.69(5)	1.61(3)	1.49(4)	1.74(6)	1.84(5)
1.84(15)		[a]1.73(1)	1.72(3)	1.61(7)	1.50(15)	[a]1.69(1)	1.86(10)
[c]1.89(5)			1.74(3)	1.65(6)	1.55(3)		1.96(7)
1.94(7)			1.75(4)	1.66(6)	1.67(7)		1.97(6)
1.85(8)			1.75(10)	[a]1.671(4)	1.70(4)		[a]1.71(1)
			1.86(6)	[a]1.693(4)	[a]1.724(12)		
			1.72(4)	1.70(5)	[a]1.734(4)		
				1.71(3)	1.82(7)		
				[a]1.750(4)	1.83(9)		
				1.80(13)	[a]1.72(1)		
				1.83(3)			
				[a]1.70(1)			

Zr	Nb	Mo	Tc	Ru	Rh	Pd	Ag
[c]2.21(4)	1.91(3)	1.52(3)		1.69(7)	1.7(1)		2.19(5)
	2.0(1)	1.58(10)		1.69(9)	[a]1.765(11)		
	1.92(4)	1.68(5)		1.74(11)	1.77(6)		
		1.68(6)		1.75(10)	[a]1.808(12)		
		1.73(2)		1.76(6)	1.85(5)		
		[a]1.860(4)		1.76(7)	2.1(1)		
		2.02(9)		1.76(10)	[a]1.79(2)		
		2.19(6)		[a]1.773(2)			
		[a]1.85(1)		[a]1.792(5)			
				1.81(2)			
				[a]1.821(3)			
				1.85(4)			
				1.85(10)			
				1.91(3)			
				1.95(7)			
				[a]1.79(1)			

Hf	Ta	W	Re	Os	Ir	Pt	Au
[a]2.069(7)		1.86(6)	1.77(10)	1.6(1)	1.73(10)	1.78(10)	
[a]2.120(8)		[a]1.870(4)	[a]1.81(2)	[a]1.754(8)	[a]1.75(3)		
[a]2.13(1)		[a]1.875(4)	[a]1.878(7)	1.79(17)	1.94(7)		
[a]2.10(1)		[a]1.876(6)	[a]1.87(1)	1.80(2)	[a]1.78(5)		
		[a]1.894(12)		[a]1.808(10)			
		[a]1.897(5)		1.81(2)			
		[a]1.88(1)		[a]1.813(4)			
				[a]1.819(3)			
				[a]1.824(13)			
				[a]1.834(11)			
				[a]1.845(3)			
				[a]1.85(10)			
				[a]1.857(4)			
				1.87(6)			
				1.87(7)			
				[a]1.883(9)			
				2.03(14)			
				[a]1.83(1)			

Table 3. Terminal M−H Bonds

Complex[a]	M−H (in Å)[b]	type[c]	Ref.
$HZr(C_8H_{11})(Me_2PCH_2CH_2PMe_2)_2$	1.67	X	395
$HV(CO)_4(Ph_2PCH_2CH_2PPh_2)$	1.55(9)	X	336
$HNbCp_2(CO)$	$(1.5)^d$	X	259
H_3NbCp_2	1.69(4)	X	54
$[HNbCp(C_5H_4)]_2$	1.70(3)	X	57
$\{HNb[C_5H_3(SiMe_2)O(SiMe_2)C_5H_4]\}_2^{2-}$ Na_2^+	1.62	X	260
$H_2Nb(Cp)(PPh_3)_2(CO)$	−	X	261
H_3TaCp_2	1.774(3)	N	54
$HTa(CO)_2(Me_2PC_2H_4PMe_2)_2$	−	X	262
H_2MoCp_2	1.685(3)	N	39, 56
$HMoCp_2[Mg(THF)_2Br]$	−	X	263
$HMo[P(OMe)_3]_4(O_2CCF_3)$	1.59(4)	X	396
$[HMoCp_2Li]_4$	−	X	166, 167
$[HMoCp_2(Mg_2Br_2R \cdot Et_2O)]_2$ $(R = Cy, Pr^i)$	−	X	264
$[HMo(C_2H_4)_2(Ph_2PC_2H_2PPh_2)_2]^+[CF_3CO_2]^-$	−	X	265
$H_4Mo(PMePh_2)_4$	1.70(3)	X	266
$H_4Mo_2(PMe_3)_6$	1.82(4)	X	267
$[HWCp_2Li]_4$	−	X	167
$HW(CO)_3Cp$	−	X	1
$H_6W(PPhPr_2^i)_3$	1.728(7)	X,N	268
$[HWCp(C_5H_4)]_2$	−	X	58
$[H_3W_2Cp_4]^+[ClO_4]^-$	$(1.604)^e$	X	38
$HW_2Cp_2(C_5H_4)_2(CH_2SiMe_3)$ (cis)	−	X	58
$HW_2Cp_2(C_5H_4)_2(CH_2SiMe_3)$ (trans)	−	X	58
$H_4W_4(CO)_{12}(OH)_4 \cdot 4\ Ph_2EtPO$	−	X	269
$HMnCp(CO)_2(SiPh_3)$	1.55	X	161, 162
$HMnCp(CO)_2(SiCl_2Ph)$	−	X	161
$HMn(CO)_5$	1.601(16)	N	45
$HMn(CO)_5$	1.576(18)	E	46
$HRe(Cp)(CO)_2(SiPh_3)$	1.66	X	161
$HReCp(CO)_2(CH_2Ph)$	−	X	71
$[H_2Re(CO)_4]^-[NEt_4]^+$ (cis)	$(1.75)^d$	X	49
$[H_2Re(CO)_4]^-[NEt_4]^+$ (trans)	−	X	50
$H_2Re(NO)(PPh_3)_3$	1.65^e	X	35
$H_3Re(Ph_2PC_2H_4PPh_2)_2$	−	X	33
$H_3Re(Ph_2PC_2H_4PPh_2)(PPh_3)_2$	−	X	33
$H_5Re(PPh_3)_3$	−	X	270
$H_5Re(PMe_2Ph)_3$	−	X	271
$H_4Re_2(CO)_6(SiEt_2)_2$	−	X	154
$H_2Re_2(CO)_7(SiEt_2)_2$	−	X	157
$[H_9Re]K_2$	1.68(1)	N	44
$H_7Re(PMe_2Ph)_2$	−	X	51
$H_8Re_2(PEt_2Ph)_4$	1.669(7) A	N	60
$HReOs_3(CO)_{15}$	−	X	272
$[H_4Re_4(CO)_{15}]^{2-}[NEt_4]_2^+$	−	X	79
$HRe_6Cl_6(CH_2SiMe_3)_9$	−	X	273
$HFe(CO)Cp(SiCl_3)_2$	−	X	274
$HFe(CO)Cp(SiF_2Me)_2$	1.49(6)	X	163
$HFe(CO)Cp(SiMe_2Ph)_2$	−	X	163
$[HFe(CO)_4]^-[(Ph_3P)_2N]^+$	1.57(12)	X	48
$H_2Fe(CO)_4$	1.556(21)	E	46

Table 3   (continued)

Complex[a]	M–H (in Å)[b]	type[c]	Ref.
$H_2Fe[P(OEt)_2Ph]_4$ (cis)	1.51(4)	X	47
$H_2Ru[PPh(OEt)_2]_4$ (trans)	1.60(8)	X	275
$[HRu(CNBu^t)_5]^+[Me_2C_2B_4H_5]^-$	–	X	276
Sr_2RuD_6	1.69(4)[f]	N[f]	83
$HRu(NO)(PPh_3)_3$	–	X	277
$H_2Ru(PPh_3)_2(C_2B_9H_{11})$	–	X	278
$HRuCl(PPh_3)_3$	1.70(15)	X	279
$HRu(C_{10}H_7)(Me_2PC_2H_4PMe_2)_2$	1.7	X	280
$HRu(CO)(PPh_3)_2(MeC_6H_4N_3C_6H_4Me)$	–	X	281
$HRu(PPh_3)_3[CH=C(Me)C(O)OC_4H_9]$	–	X	77
$[HRu(PPh_3)_2(\eta^6\text{-Ph-PPh}_2)]^+[BF_4]^-$	1.7	X	75
$\{HRu[CH_2P(Me)C_2H_4PMe_2](Me_2PC_2H_4PMe_2)\}_2$	1.67(7)	X	76
$[HRu(C_4H_6)(PMe_2Ph)_3]^+[PF_6]^-$	–	X	74
$HRuCl(C_8H_{12})(NC_5H_{11})_2$	1.57(4)	X	282
$H_2Ru(PPh_3)_4$	–	X	283
$H_2Ru(PPh_3)_3(N=N-B_{10}H_8\text{-}SMe_2)$	1.63(8)	X	284
$HRu(CO)(PPh_3)_2(MeC_6H_4NNNC_6H_4Me)$	1.8	X	285
$HRu(CO)(PPh_3)_2(MeC_6H_4NCHNC_6H_4Me)$	1.58(7)	X	286
$[HRu(PMe_2Ph)_5]^+[PF_6]^-$	–	X	287
$[HRu(C_8H_{12})(NH_2NMe_2)_3]^+[PF_6]^-$	1.36	X	288
$HRu(PPh_3)_3(MeCOO)$	1.68	X	289
$HRu(PPh_3)_3(HCOO)$	–	X	290
$H_4Ru_2(PMe_3)_6$	1.68(6)	X	291
$HOs(CO)(PPh_3)_2(N_2Ph)$	–	X	292
$HOs(C_{10}H_7)(Me_2PC_2H_4PMe_2)_2$	–	X	280b
$HOsBr(CO)(PPh_3)_3$	–	X	293
$HOs(CSSMe)(CO)_2(PPh_3)_2$	1.64(6)	X	6
$H_4Os(PMe_2Ph)_3$	1.659(3)	N	59
$H_4Os(PEt_2Ph)_3$	–	X	294
$H_2Os_3(CO)_{11}$	–	X	78
$H_2Os_3(CO)_{10}(PPh_3)$	1.52(7)	X	209
$H_2Os_3(CO)_{10}(CNBu^t)$	–	X	205
$HOs_3(CO)_{10}(PEt_3)(CF_3CCHCF_3)$	–	X	295
$HCo(CO)(PPh_3)_3$	1.41(9)	X	296
$HCo(BH_4)(PCy_3)_2$	1.34(8)	X	93
$HCo[PPh(OEt)_2]_4$	1.38(16)	X	297
$HCo(PF_3)_4$	–	X	61
$HCo(CO)_4$	1.556(18)	E	46
$HCo[P(C_6H_4PPh_2)_3]$	1.60(16)	X	64
$HCo[P(C_2H_4PPh_2)_3]$	1.43(6)	X	65
$[HCoP(C_2H_4PPh_2)_3]^+[BF_4]^-$	1.53(15)	X	66
$HCo[N(C_2H_4PPh_2)_3]$	1.45(5)	X	67
$HCo[N(C_2H_4PPh_2)_3]\cdot THF$	1.38(10)	X	68
$HCo(N_2)(PPh_3)_3$	1.65(12)	X	298
$HRh(N_2)(PPhBu^t_2)_2$	1.66(5)	X	299
$HRh(PPh_3)_2(C_2B_9H_{11})$	1.54(9)	X[f]	300
$HRh(PPh_3)(C_2B_9H_{10}\text{-}C_2H_4CH\text{-}CH_2)$	1.65(8)	X	72
$HRh(PPh_3)_3$	–	X	301
Li_4RhH_4	(1.9)[d]	X	84
$HRh(PPh_3)_2(SiCl_3)Cl$	1.48	X	302
Li_4RhH_5	(1.95)[d]	X	84

Table 3 (continued)

Complex[a]	M–H (in Å)[b]	type[c]	Ref.
HRh(CO)(PPh$_3$)$_3$	1.60(12)	X	17
HRh(PPh$_3$)$_4$	–	X	62
HRh(PPh$_3$)$_3$(AsPh$_3$)	–	X	63
H$_2$RhCl(PBu$_3^t$)$_2$	1.36(8)	X	303
[HRh(NH$_3$)$_5$]$^+$[ClO$_4$]$^-$	1.82(17)	X	304
[HRhCl(Ph$_2$PCH$_2$PPh$_2$)$_2$]$^+$[BPh$_4$]$^-$	1.51(7)	X	305
HRhCl(Bu$_2^t$PCH$_2$CH$_2$CHCH$_2$CH$_2$PBu$_2^t$)	1.36(3)	X	306
Sr$_2$IrD$_5$	1.70(4)[f]	N[f]	83
HIr(CO)$_2$(PPh$_3$)$_2$	1.64(5)	X	307
[HIr(NO)(PPh$_3$)$_3$]$^+$[ClO$_4$]$^-$	–	X	308
HIr(PPr$_3^i$)$_2$(C$_4$H$_6$)	1.77(12)	X	309
HIrCl(η^3-C$_3$H$_4$Ph)(PPh$_3$)$_2$	1.57(7)	X	73
HIr(CO)(PPh$_3$)$_2$(NCCH=CHCN)	–	X	310
HIrCl$_2$(Me$_2$SO)$_3$	–	X	1
[HIr(COOMe)(Me$_2$PC$_2$H$_4$PMe$_2$)$_2$]$^+$[BPh$_4$]$^-$	–	X	1
H$_2$Ir(CO)(GeMe$_3$)(PPh$_3$)$_2$	–	X	311
HIrI(PPh$_3$)$_2$(MeOC$_6$H$_3$NNH)	–	X	312
mer-H$_3$Ir(PPh$_3$)$_3$	1.60	X	52
fac-H$_3$Ir(PMe$_2$Ph)$_3$	–	X	51
[H$_2$Ir(CO)(PPh$_3$)$_3$]$^+$[SiF$_5$]$^-$	–	X	313
{[HIr(PPh$_3$)$_2$]$_2$(μ-Cl)(μ-SPh)$_2$}$^+$[ClO$_4$]$^-$	–	X	314
HIrBr(Ph)(CO)(PEt$_3$)$_2$	–	X	315
HIr(PPh$_2$C$_6$H$_4$)$_2$(PPh$_3$)	1.82(17)	X	316
HIrI(PPh$_3$)$_2$(MeOC$_6$H$_3$NNH)	–	X	317
[H$_2$Ir(Ph$_2$PCH$_2$CH$_2$PPh$_2$)$_2$]$^+$[PF$_6$]$^-$	–	X	318
[H$_2$Ir(Ph$_2$PCH$_2$CH$_2$PPh$_2$)$_2$]$^+$[B$_9$H$_{14}$]$^-$	1.77	X	319
HIr(CO)(PPh$_3$)$_2$(Me$_2$SiOSiMe$_2$)·EtOH	–	X	320
H$_2$IrCl[Ph$_2$P(CH$_2$)$_2$CH=CH(CH$_2$)$_2$PPh$_2$]	–	X	321
[HIr(SBu^t)(CO)(P(OMe)$_3$)]$_2$	(1.7)[d]	X	322
H$_2$Ir$_2$(CO)$_4$(PPh$_3$)$_2$(μ-SO$_2$)	–	X	323
{H$_5$Ir$_2$[Ph$_2$P(CH$_2$)$_3$PPh$_2$]$_2$}$^+$[BF$_4$]$^-$	–	X	81
{H$_7$Ir$_3$[Ph$_2$P(CH$_2$)$_3$PPh$_2$]$_3$}$^{2+}$[BF$_4$]$_2^-$	–	X	81
[H$_7$Ir$_3$(py)$_3$(PCy$_3$)$_3$]$^{2+}$[PF$_6$]$_2^-$	–	X	80
[H$_2$Ir$_4$(CO)$_{10}$]$^{2-}$[(Ph$_3$P)$_2$N]$_2^+$	–	X	324
HNiCl(PPr$_3^i$)$_2$	1.37	X	1
{HNi[N(C$_2$H$_4$PPh$_2$)$_3$]}$^+$BF$_4^-$	1.57(20)	X	69
HNi(BH$_4$)(PCy$_3$)$_2$	1.48(7)	X	94
HPdCl(PEt$_3$)$_2$	–	X	325
HPdCl(PPr$_3^i$)$_2$	–	X	1
[HPt(PPh$_3$)$_3$]$^+$[(CF$_3$CO$_2$)$_2$H]$^-$	–	X	326
HPtBr(PEt$_3$)$_2$	–	X	327
HPtCl(PEtPh$_2$)$_2$	–	X	328
HPt(B$_9$H$_{10}$S)(PEt$_3$)$_2$	1.66	X	329
HPt(CS$_2$H)(PCy$_3$)$_2$	–	X	330
HPt(MeC$_6$H$_4$NNNC$_6$H$_4$Me)(PPh$_3$)$_2$	–	X	331
[HPt(PhHNNCMe$_2$)(PPh$_3$)$_2$]$^+$[BF$_4$]$^-$	–	X	6
HPt(CH$_2$CN)(PPh$_3$)$_2$	–	X	70
HPt(SiH$_3$)(PCy$_3$)$_2$	–	X	332
H$_2$Pt(PCy$_3$)$_2$ (trans)	–	X	333
HPt(PEt$_3$)$_2$(C$_2$B$_4$H$_7$)	–	X	276, 334
[HPt$_2$(Ph$_2$PCH$_2$PPh$_2$)$_3$]$^+$[PF$_6$]$^-$	–	X	335

Table 3   (continued)

Complex[a]	M–H (in Å)[b]	type[c]	Ref.
$[H_2Pt_2(Ph)(PEt_3)_4]^+[BPh_4]^-$	–	X	176
$\{H_3Pt_2[Bu_2^tP(CH_2)_3PBu_2^t]_2\}^+[BPh_4]^-$	–	X	177
$[HZnN(Me)C_2H_4NMe_2]_2$	1.62	N	165
$H_4Th_2(C_5Me_5)_4$	2.03(1)	N	55

[a] In this table we have adopted the convention (for ionic compounds) of listing the hydrido species first

[b] Whenever appropriate, averages are reported. Numbers in parentheses give the estimated standard deviation of the last reported digit. No estimated error is given in cases when hydrogen atom positions were not varied in least-squares calculations, or when no standard deviation is reported

[c] N = neutron diffraction, X = X-ray diffraction, E = electron diffraction

[d] Distances derived from assumed H positions

[e] Distances derived from calculated H positions

[f] Distances derived from powder neutron diffraction data

Table 4. Compounds with M–H–X Bonds & Suspected M … H–X Interactions

Complex[a]	n[b]	M	X	M–H (Å)[c]	M–X (Å)[c]	M–H–M (°)[c]	Type[d]	Ref.
$Sc(BH_4)_3(THF)_2$	3	Sc	B	2.09(5)	2.280(5)	86	X	86a
	2	Sc	B	2.09(5)	2.551(5)	102	X	
$Y(BH_4)_3(THF)_3$	3	Y	B	2.36(11)	2.58(1)	91	X	86b
	2	Y	B	2.27(9)	2.68(2)	103	X	
$TiCp_2(BH_4)$	2	Ti	B	1.75(8)	2.37(1)	104(5)	X	96a
$TiCp_2(BH_4)$	2	Ti	B	1.89(5)	2.31(4)	–	E	96b
$[TiCp(BH_4)Cl]_2$	3	Ti	B	1.94(7)	2.17(1)	86	X	337
$[Ti(BH_3)_2(C_{16}H_{16}N_2O_2)]_2$	1	Ti	B	1.84(15)	–	–	X	338
$Zr(BH_4)_4$	3	Zr	B	2.21(4)	2.308(10)	80.0	E	100
$Zr(BH_4)_4$	3	Zr	B	–	2.34(3)	–	X	99
$Hf(BH_4)_4$	3	Hf	B	2.13(1)	2.28(1)	80.6(6)	N	101
$Hf(C_5H_4Me)_2(BH_4)_2$	2	Hf	B	2.069(7) 2.120(8)	2.553(6)	96.8(5)	N	28
$NbCp(BH_4)_4$	2	Nb	B	2.0(1)	2.26(6)	89	X	97
$[Cr(CO)_4(B_3H_8)]^-[NMe_4]^+$	1	Cr	B	1.78(6)	2.44(1)	104(4)	X	104
$[Mo(CO)_4(BH_4)]^-[(Ph_3P)_2N]^+$	2	Mo	B	2.02(9)	2.41(2)	95(5)	X	92
$Mo(CO)_2(\eta^3\text{-}C_7H_7)\text{-}$ $[H_2B(N_2C_3HMe_2)_2]$	1	Mo	B	2.10(6)	2.797(7)	110	X	123
$Mo(CO)_2(\eta^3\text{-}C_3H_5)\text{-}$ $[H_2B(N_2C_3HMe_2)_2]$	1	Mo	B	2.30	2.81	102	X	122
$Mn(CO)_3(B_8H_{13})$	1	Mn	B	1.74(3)	2.256(4)	99(2)	X	106
$Mn(CO)_3[B_9H_{12}O(CH_2)_4\text{-}$ $NEt_3]$	1	Mn	B	1.67(3)	2.236(7)	103	X	339
$Mn(CO)_3(B_3H_8)$	1	Mn	B	1.69(5)	2.290(7)	104(3)	X	105
$Mn(CO)_3(B_9H_{12})(2\text{-}THF)$	1	Mn	B	1.75(4)	2.237(5)	95.3	X	108
$Mn(CO)_3(B_9H_{12})(6\text{-}THF)$	1	Mn	B	1.75	2.219(5)	95.9	X	108b
$HMn_3(CO)_{10}(BH_3)_2$	1	Mn	B	–	2.30(2)	–	X	121
$[Fe(CO)_3(B_5H_8)]^-[NBu_4^n]^+$	1	Fe	B	1.52	2.13(2)	105	X	110

Table 4   (continued)

Complex[a]	n[b]	M	X	M–H (Å)[c]	M–X (Å)[c]	M–H–M (°)[c]	Type[d]	Ref.
Fe(CO)$_3$(B$_5$H$_8$)Cu(PPh$_3$)$_2$	1	Fe	B	1.56(6)	2.115(10)	–	X	113
	1	Cu	B	1.96(7)	2.164(8)	–		
HCo(PCy$_3$)$_2$(BH$_4$)	2	Co	B	1.83(9)	2.14(1)	83(4)	X	93
Co(B$_4$H$_8$)Cp	1	Co	B	1.43(5)	2.135(8)	96	X	112
5-Co(B$_9$H$_{13}$)Cp	1	Co	B	1.49(4)	2.072(6)	97.3(4)	X	109
Co(C$_2$B$_7$H$_{11}$)Cp	1	Co	B	1.40(4)	2.068(5)	99	X	340
Co[MeC(CH$_2$PPh$_2$)$_3$](BH$_4$)	2	Co	B	1.50(15)	2.21(3)	103	X	95
Co(terpyridine)(BH$_4$)	2	Co	B	1.724(12)	2.15	90.0	N	98
{Co[(B$_{10}$C$_2$H$_{10}$)$_2$]$_2$}$^-$[NEt$_4$]$^+$	1	Co	B	1.82(7)	2.29(1)	100(5)	X	117
[Rh(C$_2$B$_9$H$_{11}$)(PPh$_3$)]$_2$	1	Rh	B	1.77(6)	2.332(8)	99(5)	X	120
Rh(CB$_{10}$H$_{10}$Ph)(PPh$_3$)$_2$	1	Rh	B	2.1	2.35(3)	–	X	118
Ir(B$_5$H$_8$)(CO)(PPh$_3$)$_2$	1	Ir	B	1.73	2.250(6)	91	X	111
HNi(PCy$_3$)$_2$BH$_4$	2	Ni	B	1.74(6)	2.201(8)	92(3)	X	94
Cu(B$_3$H$_8$)(PPh$_3$)$_2$	1	Cu	B	1.84(5)	2.30(1)	99(3)	X	103
Cu(BH$_4$)(PPh$_3$)$_2$	2	Cu	B	1.82(3)[e]	2.184(9)	94[e]	X	341
Cu(BH$_4$)(PMePh$_2$)$_3$	1	Cu	B	1.697(5)	2.518(3)	121.7(4)	X,N	29, 90
[Cu(PPh$_3$)$_2$(NCBH$_3$)]$_2$	1	Cu	B	1.86(10)	2.91(2)	153(8)	X	91
[Cu(PPh$_3$)$_2$]$_2$(B$_{10}$H$_{10}$)	1	Cu	B	1.97(6)	2.30(1)	93(5)	X	115
[Ag(PPh$_3$)(B$_8$C$_2$H$_{11}$)]$_2$	1	Ag	B	2.19(5)	3.033(6)	130(3)	X	119
[U(BH$_4$)$_4$]$_n$	3	U	B	2.34(2)	2.52(1)	83(1)	N	102
	2	U	B	2.40(3)	2.86(4)	98(1)	N	
U(BH$_4$)$_4$(THF)$_2$	3	U	B	–	2.56(4)	–	X	87
[U(BH$_4$)$_4$(Pr$_2^n$O)]$_2$	3	U	B	–	2.52(5)	–	X	88
	2	U	B	–	2.87(5)	–	X	
[U(BH$_4$)$_4$(Me$_2$O)]$_n$	3	U	B	–	2.52(2)	–	X	89
	2	U	B	–	2.88(2)	–	X	
[U(BH$_4$)$_4$(Et$_2$O)]$_n$	3	U	B	–	2.54(1)	–	X	89
	2	U	B	–	2.89(1)	–	X	
[HTi(C$_5$H$_4$)AlEt$_2$]$_2$(C$_{10}$H$_8$)	1	Ti	Al	–	2.82	102	X	124
[HTiCp(C$_5$H$_4$)AlEt$_2$]$_2$	1	Ti	Al	(1.8)[f]	2.79	(101)[f]	X	125
(H)[TiCp]$_2$(H$_2$AlEt$_2$)(C$_{10}$H$_8$)	1	Ti	Al	1.69	3.134	135	X	124
{Ta[H$_2$Al(OC$_2$H$_4$OMe)$_2$]- (Me$_2$PC$_2$H$_4$PMe$_2$)$_2$}$_2$	2	Ta	Al	–	2.733(9)	–	X	127
[HMoCp(C$_5$H$_4$)]$_2$Al$_3$Me$_5$	1	Mo	Al	–	2.973(5)	–	X	126
Cp$_2$Zr[CH$_2$CH(AlEt$_2$)$_2$]Cp	1	Zr	C	2.64(6)	2.393(4)	–	X	144
				2.64(6)	2.880(6)	–		
Ta(=CHCMe$_3$)(PMe$_3$)- (C$_2$H$_4$)Cp	1	Ta	C	2.042(5)	1.946(3)	70.1(3)	N	342
[Ta(=CHCMe$_3$)(PMe$_3$)Cl$_3$]$_2$	1	Ta	C	2.119(4)	1.898(2)	63.1(2)	N	148
Cr$_2$(PMe$_3$)$_2$(CH$_2$SiMe$_3$)$_4$	1	Cr	C	2.3[g]	–	–	X	142
Mo(CO)$_2$(η^3-C$_7$H$_7$)[Et$_2$B- (N$_2$C$_3$H$_3$)$_2$]	1	Mo	C	1.93[g]	2.92(2)	–	X	131
Mo(CO)$_2$(η^3-C$_3$H$_4$Ph)- [Et$_2$B(N$_2$C$_3$H$_3$)$_2$]	1	Mo	C	2.27(8)	3.055(7)	136(6)	X	130
MnCp(C$_7$H$_8$)(CO)$_2$	1	Mn	C	–	3.321(8)	–	X	146
[Fe(P(OMe)$_3$)$_3$(η^3-C$_8$H$_{13}$)]$^+$- [BF$_4$]$^-$	1	Fe	C	1.874(3)	2.362(2)	99.4(2)	N	133
HFe$_4$(η^2-CH)(CO)$_{12}$	1	Fe	C	1.750(4)	1.924(2)	79.6(5)	X,N	151
Ru(PPh$_3$)$_3$Cl$_2$	1	Ru	C	2.59[g]	–	–	X	134
HRu(PPh$_3$)$_3$Cl	1	Ru	C	2.85[g]	–	–	X	136

Table 4   (continued)

Complex[a]	n^b	M	X	M–H (Å)[c]	M–X (Å)[c]	M–H–M (°)[c]	Type[d]	Ref.
RuCl[CN(C_6H_4Me)CH_2CH_2- NC_6H_3Me](PEt$_3$)$_2$	1	Ru	C	2.23[g]	–	–	X	143
[Rh(PPh$_3$)$_3$]$^+$[ClO$_4$]$^-$	1	Rh	C	2.56[g]	2.62(2)	–	X	140
Rh(PPh$_3$)$_3$Cl	1	Rh	C	2.77[g]	–	–	X	138
				2.84[g]	–	–		
HRh(PPh$_3$)$_2$(SiCl$_3$)Cl	1	Rh	C	2.79[g]	–	–	X	137
PdBr(PPh$_3$)$_2$[C$_4$(COOMe)$_4$H]	1	Pd	C	2.3[g]	–	–	X	129
Pd(PPhBu$_2^t$)$_2$	1	Pd	C	2.70(10)	–	–	X	139
				2.83(10)				
Pd(PMe$_2$Ph)$_2$I$_2$	1	Pd	C	2.8[g]	–	–	X	135
Pd(C$_8$H$_9$N)$_2$Cl$_2$	1	Pd	C	2.46[g]	–	–	X	145
Pt(PPhBu$_2^t$)$_2$	1	Pt	C	2.77(11)	–	–	X	139
				2.83(12)	–	–		
PtCl$_2$(PMe$_2$Ph)[C(OEt)CH_2Ph]	1	Pt	C	2.6(1)	–	–	X	141
[Cu(dien)(C$_7$H$_{10}$)]$^+$[BPh$_4$]$^-$	1	Cu	C	2.01(15)	2.78(1)	158(17)	X	147
H$_2$W$_2$(CO)$_8$(SiEt$_2$)$_2$	1	W	Si	–	2.703(4)	–	X	156
H$_2$Re$_2$(CO)$_8$(SiPh$_2$)	1	Re	Si	–	2.544(9)	–	X	155
[Pt(SiEt$_3$)(PCy$_3$)(μ-H)]$_2$	1	Pt	Si	–	2.336(11)	–	X	159
[Pt(μ-SiMe$_2$)(μ-H)(PCy$_3$)]$_2$	1	Pt	Si	1.78	2.420(2)	87.6	X	158
MoCp$_2$(μ-H)$_2$(ZnBr$_2$ · DMF)	2	Mo	Zn	1.58	2.793(3)	118	X	164

a   In this table we have adopted the convention (for ionic compounds) of listing the hydrido species first

b   1 = M(μ-H)X, 2 = M(μ-H)$_2$X, 3 = M(μ-H)$_3$X

c   Whenever appropriate, averages are reported. Numbers in parentheses give the estimated standard deviation of the last reported digit. No estimated error is given in cases when hydrogen atom positions were not varied in least-squares calculations, or when no standard deviation is reported

d   N = neutron, X = X-ray, E = electron diffraction

e   The Cu–H value for Cu(BH$_4$)(PPh$_3$)$_2$ was originally reported as 2.02(5) Å in reference 341, but was later revised to 1.82(3) Å (see footnote 27 of reference 115, and figure 3b of reference 90b)

f   Values derived from assumed hydrogen position

g   In most cases involving suspected M ... H–C interactions, the M ... H distances are based on estimated or calculated H positions

Table 5. Compounds with M−H−M bonds

Complex[a]	n[b]	M	M−H (Å)[c]	M−M (Å)[c]	M−H−M (°)[c]	Type[d]	Ref.
$HTi_2Cp_2(H_2AlEt_2)(C_{10}H_8)$	1	Ti	1.80	3.374	138	X	124
$HZr_2Cp_4(C_{10}H_7)$	1	Zr	−	3.307(2)	−	X	343
$NbCp_2(CO)(H)Fe(CO)_4$	1	Nb	1.91(3)	3.318(1)	141(2)	X	178
		Fe	1.61(3)				
$[HCr_2(CO)_{10}]^-[(Ph_3P)_2N]^+$	1	Cr	1.675	3.349(13)	(180?)[e]	N	171
$[DCr_2(CO)_{10}]^-[(Ph_3P)_2N]^+$	1	Cr	1.734(9)	3.390(3)	155.7(9)	N	172
$[HCr_2(CO)_{10}]^-[NEt_4]^+$	1	Cr	1.722(20)	3.386(6)	158.9(6)	X, N	169, 170
$HCrRe(CO)_{10}$	1	Cr, Re	−	3.435(1)	−	X	344
$H[MoCp(CO)_2]_2(PMe_2)$	1	Mo	1.860(4)	3.267(2)	122.9(2)	X, N	21
$[HMo_2(CO)_{10}]^-[(Ph_3P)_2N]^+$	1	Mo	1.80	3.4219(9)	136	X	175b
$[HMo_2(CO)_{10}]^-[K(crypt-222)]^+$	1	Mo	1.90	3.4056(5)	127	X	175b
$[HMo_2(CO)_9(PPh_3)]^-[NEt_4]^+$	1	Mo	1.68(5) 2.19(6)	3.474(1)	127(3)	X	175a
$[HMo_2Cl_8]^{3-}[C_5H_6N]_3^+$	1	Mo	1.68(6)	2.371(1)	89.6(3)	X	196
$[HMo_2Cl_8]^{3-}[Mo(O)Cl_4(H_2O)]^-$- $[H_5O_2]^+[NEt_4]_3^+$	1	Mo	1.73(2)	2.375(2)	86.8(1)	X	197
$H_4Mo_2(PMe_3)_4$	2	Mo	1.52(3)	2.194(3)	92	X	267
$[H(MoCp)_2(OH)(C_{10}H_8)]^{2+}$- $[PF_6]_2^-$	1	Mo	−	3.053	−	X	345
$[H_2W_2(CO)_8]^{2-}[NEt_4]_2^+$	2	W	1.86(6)	3.016(1)	109(5)	X	180
$HW_2(CO)_9NO$ (triclinic form)	1	W	1.875(4)	3.328(3)	125.0(2)	X, N	23
$HW_2(CO)_9NO$ (monoclinic form)	1	W	1.870(4)	3.330(3)	125.9(4)	N	23
$HW_2(CO)_8(NO)[P(OMe)_3]$	1	W	1.876(6)	3.393(4)	129.4(3)	N	168
$[HW_2(CO)_{10}]^-[NEt_4]^+$	1	W	1.894(12)	3.528(2)	137.1(10)	X, N	4, 173
$[HW2(CO)_{10}]^-[(Ph_3P)_2N]^+$	1	W	−	3.391(1)	−	X	173
$[HW_2(CO)_{10}]^-[PPh_4]^+$	1	W	1.897(5)	3.340(5)	123.4(5)	N	4b, 174
$[H_3W_2(C_5H_5)_4]^+[ClO_4]^-$	1	W	$(1.856)^f$	3.628	$(155)^f$	X	38
$H_2W_2(OPr^i)_{14}$	1	W	−	2.446(1)	−	X	346
$[Cp_2W(\mu-H)_2Rh(PPh_3)_2]^+[PF_6]^-$	2	W, Rh	−	2.721(8)	−	X	183
$HMn_2(CO)_8(PPh_2)$	1	Mn	1.86(6)	2.937(5)	104(5)	X	347
$H_3Mn_3(CO)_{12}$	1	Mn	1.72(3)	3.111(2)	131(7)	X	18
$HMn_3(CO)_{10}(BH_3)_2$	1	Mn	1.65(10)	2.845(3)	−	X	121
$HMnRe_2(CO)_{14}$	1	Re	−	3.392(2)	−	X	348
$H_2Re_2(CO)_8$	2	Re	−	2.896(3)	−	X	179
$H_8Re_2(PEt_2Ph)_4$	4	Re	1.878(7)	2.538(4)	85.0(3)	N	60
$HRe_3(CO)_{14}$	1	Re	−	3.295(2)	−	X	1
$HRe_3(CO)_{12}(SnMe_2)$	1	Re	−	3.23	−	X	349
$[H_2Re_3(CO)_{12}]^-[AsPh_4]^+$	1	Re	−	3.177(5)	−	X	11
$[HRe_3(CO)_{12}]^{2-}[NEt_4]_2^+$	1	Re	−	3.125(3)	−	X	50
$[HRe_3(CO)_{12}]^{2-}[AsPh_4]_2^+$	1	Re	−	3.144(5)	−	X	350
$[H_4Re_3(CO)_{10}]^-[NEt_4]^+$	2	Re	−	2.821(7)	−	X	186
	1	Re	−	3.184(7)	−		
$[H_3Re_3(CO)_{10}]^{2-}[NEt_4]_2^+$	2	Re	−	2.797(4)	−	X	185a
	1	Re	−	3.031(5)	−		
$H_3Re_3(CO)_{10}(py)_2$	1	Re	−	3.292	−	X	351
$[H_3Re_3(CO)_9O]^{2-}[NEt_4]_2^+$	1	Re	−	2.968(1)	−	X	185
$H_3Re_3(CO)_8[(EtO)_2POP(OEt)_2]_2$	1	Re	1.81(2)	3.282(5)	130(1)	N	202
$[H_4Re_4(CO)_{15}I]^-[NEt_4]^+$	1	Re	−	3.282(1)	−	X	352

Table 5 (continued)

Complex[a]	n[b]	M	M–H (Å)[c]	M–M (Å)[c]	M–H–M (°)[c]	Type[d]	Ref.
$[H_4Re_4(CO)_{15}]^{2-}[NEt_4]_2^+$	1	Re	–	3.287(2)	–	X	79
	1	Re	–	3.192(2)	–		
$[H_4Re_4(CO)_{13}]^{2-}[NEt_4]_2^+$	1	Re	–	3.094(4)	–	X	36
$[H_6Re_4(CO)_{12}]^{2-}[NMe_3Bz]_2^+$	1	Re	–	3.157(6)	–	X	218
$[H_6Re_4(CO)_{12}]^{2-}[AsPh_4]_2^+$	1	Re	–	3.160(7)	–	X	14
$\{H_3Fe_2[MeC(CH_2PPh_2)_3]_2\}^+ \cdot$							
$[PF_6]^-$	3	Fe	1.83(3)	2.332(3)	79.3	X	187
$[HFe_2(CO)_8]^-[(Ph_3P)_2N]^+$	1	Fe	1.61(7)	2.521(1)	103(4)	X	353
$[HFe_3(CO)_{11}]^-[NHEt_3]^+$	1	Fe	–	2.577(3)	–	X	354
$[HFe_3(CO)_{11}]^-[(Ph_3P)_2N]^+$	1	Fe	1.693(4)	2.589(2)	99.7(2)	N	355
$HFe_3(CO)_9(SPr^i)$	1	Fe	1.80(13)	2.678(2)	96(6)	X	15
$HFe_3(CO)_{10}(COMe)$	1	Fe	–	–	–	X	356
$HFe_3(CO)_9(N=CHMe)$	1	Fe	1.65(6)	2.588(1)	103(3)	X	219
$HFe_3(CO)_9(HN=CMe)$	1	Fe	1.71(3)	2.754(1)	107(2)	X	219
$HFe_4(\eta^2\text{-}CH)(CO)_{12}$	1	Fe	1.671(4)	2.599(2)	102.1(2)	X, N	151
$[HFe_4(CO)_{13}]^-[NMe_3Bz]^+$	1	Fe	1.66(6)	2.619(1)	104(3)	X	152
$HFe_5N(CO)_{14}$	1	Fe	1.70(5)	2.6024(7)	103(3)	X	227
$H_2FeRu_3(CO)_{13}$	1	Ru	–	2.905(6)	–	X	217
$[HFeRu_3(CO)_{13}]^-[(Ph_3P)_2N]^+$	1	Ru	1.821(3)	2.916(2)	106.4(2)	N	222
$Ru_2(H)(\mu\text{-}H)(Cl)(\mu\text{-}Cl)\cdot$							
$(C_8H_{12})_2(NH_2NMe_2)$	1	Ru	1.75	2.91(1)	112.3	X	357
$HRu_2(CO)_3[P(OC_6H_4)(OPh)_2]_2\cdot$							
$[OP(OPh)_2]$	1	Ru	–	2.889(6)	–	X	358
$HRu_2(CO)_6[PhC=CH\text{-}C(O)Me]$	1	Ru	1.74(11)	2.862(3)	110(6)	X	359
$H_4Ru_2(PMe_3)_4$	2	Ru	1.95(7)	2.811(4)	92	X	291
$[H_3Ru_2(PMe_3)_6]^+[BF_4]^-$	3	Ru	1.91(3)	2.540(1)	83	X	291
$HRu_3(CO)_9(PhCC_6H_4)$	1	Ru	–	2.914(4)	–	X	360
$HRu_3(CO)_{10}(SCH_2COOH)$	1	Ru	–	2.839(4)	–	X	361
$HRu_3(CO)_9(C_7H_4NS_2)$	1	Ru	1.85	2.836(5)	–	X	361
$HRu_3(CO)_{10}(CNMe_2)$	1	Ru	1.85(4)	2.801(1)	98(2)	X	16, 362
$HRu_3(CO)_{10}COMe$	1	Ru	1.76(7)	2.803(2)	105(4)	X	363
$HRu_3(CO)_9(\text{-}C\equiv C\text{-}Bu^t)$	1	Ru	1.792(5)	2.792(3)	102.3(2)	N	364
$HRu_3(CO)_9(C_6H_9)$	1	Ru	1.69(7)	2.994(1)	125(4)	X	365
$[HRu_3(CO)_{11}]^-[(Ph_3P)_2N]^+$	1	Ru	1.69(9)	2.815(2)	113(6)	X	390
$HRu_3(CO)_7(C_6H_9)(C_6H_{10})$	1	Ru	(1.5)	2.841(2)	–	X	387
$H_2Ru_3(CO)_9S$	1	Ru	–	2.881	–	X	366
$H_3Ru_3(CO)_9(CMe)$	1	Ru	1.81(2)[g]	2.842(6)	103(1)[g]	X[g]	40
$H_2Ru_4(CO)_{13}$	1	Ru	–	2.930(7)	–	X	216
$[H_3Ru_4(CO)_{12}]^-[(Ph_3P)_2N]^+\cdot$							
(C_2 isomer)	1	Ru	–	2.923(1)	–	X	213
$[H_3Ru_4(CO)_{12}]^-[(Ph_3P)_2N]^+\cdot$							
(C_{3v} isomer)	1	Ru	–	2.937(1)	–	X	213
$H_4Ru_4(CO)_{12}$	1	Ru	1.76	2.950(1)	–	X	210
$H_4Ru_4(CO)_{10}(PPh_3)_2$	1	Ru	–	2.966(2)	–	X	210, 211
$H_4Ru_4(CO)_{10}(Ph_2PC_2H_4PPh_2)$	1	Ru	1.76(6)	2.97(4)	115(3)	X	212
$H_4Ru_4(CO)_8[P(OMe)_3]_4$	1	Ru	1.773(2)	2.978(4)	114.2(3)	N	202
$HOs_3(CO)_{10}Cl$	1	Os	(1.85)[f]	2.846(1)	(101)[f]	X	193
$HOs_3(CO)_{10}Br$	1	Os	2.03(14)	2.851(1)	89(6)	X	194
$HOs_3(CO)_{10}(OMe)$	1	Os	–	2.863	–	X	190

Table 5   (continued)

Complex[a]	n[b]	M	M–H (Å)[c]	M–M (Å)[c]	M–H–M (°)[c]	Type[d]	Ref.
HOs$_3$(CO)$_{10}$(SEt)	1	Os	–	2.863(2)	–	X	203
HOs$_3$(CO)$_{10}$(C$_2$H$_3$)	1	Os	1.813(4) 1.857(4)	2.845(2)	101.6(2)	N	30
HOs$_3$(CO)$_{10}$(CHCH=NEt$_2$)	1	Os	1.79(17)	2.785(2)	102(8)	X	372
HOs$_3$(CO)$_{10}$(PhCNCH$_3$)	1	Os	–	2.918(1)	–	X	384
HOs$_3$(CO)$_{10}$[C=N(H)(Bu^t)]	1	Os	–	2.812(1)	–	X	385
HOs$_3$(CO)$_{10}$(CF$_3$CHCCF$_3$)	1	Os	–	2.78	–	X	374
HOs$_3$(CO)$_{10}$(C$_2$H$_3$PMe$_2$Ph)	1	Os	1.87(7)	2.800(1)	97(3)	X	369
HOs$_3$(CO)$_{10}$(NCHCF$_3$)	1	Os	1.824(13)	2.815(4)	101.0(6)	N	202
HOs$_3$(CO)$_9$(HC=NPh)	1	Os	–	2.956(1)	–	X	386
HOs$_3$(CO)$_9$(HC=NPh)-[P(OMe)$_3$]	1	Os	–	2.961(1)	–	X	386
HOs$_3$(CO)$_9$(C$_2$H$_4$)(SMe)	1	Os	–	2.842(1)	–	X	373
HOs$_3$(CO)$_9$(PPh$_3$)(PPh$_2$C$_6$H$_4$)	1	Os	–	3.047	–	X	367
HOs$_3$(CO)$_9$(C$_9$H$_{11}$O)	1	Os	–	2.965(1)	–	X	388
HOs$_3$(CO)$_8$(C$_9$H$_{11}$O)	1	Os	1.6	3.007(1)	138	X	371
H$_2$Os$_3$(CO)$_{11}$	1	Os	–	2.989(1)	–	X	78
H$_2$Os$_3$(CO)$_{10}$	2	Os	1.845(3)	2.683(1)	94.3(1)	N	30, 184
H$_2$Os$_3$(CO)$_{10}$(CH$_2$)	1	Os	1.883(9) 1.754(8)	3.053(3)	114.1(5)	N	208
	1	Os	1.834(11) 1.808(10)	2.824(3)	101.7(5)	N	
HDOs$_3$(CO)$_{10}$(CHD)	1	Os	1.85(10)	3.066(6)	112	N	25
H$_2$Os$_3$(CO)$_{10}$(PPh$_3$)	1	Os	1.87(6)	3.018(1)	107(3)	X	209
H$_2$Os$_3$(CO)$_{10}$(CNBu^t)	1	Os	–	2.930(1)	–	X	205
H$_2$Os$_3$(CO)$_9$S	1	Os	1.819(3)	2.915(1)	106.5(1)	N	207
H$_2$Os$_3$(CO)$_9$(CCH$_2$)	1	Os	–	2.905	–	X	368
H$_2$Os$_3$(CO)$_9$(PPh$_3$)	2	Os	–	2.683(2)	–	X	206
H$_2$Os$_3$(CO)$_9$(CNBu^t)	2	Os	–	2.690(1)	–	X	205
H$_2$Os$_3$(CO)$_9$(OC$_6$H$_3$CH$_2$Ph)	1	Os	–	2.949(1)	–	X	375
	1	Os	–	2.786(1)			
H$_3$Os$_3$Co(CO)$_{12}$	1	Os	–	2.901(1)	–	X	381
HOs$_3$W(CO)$_{12}$(C$_5$H$_5$)	1	Os	–	2.933(3)	–	X	221, 337
H$_3$Os$_3$W(CO)$_{11}$(C$_5$H$_5$)	1	Os	–	2.941(2)	–	X	220, 377
	1	Os, W	–	3.078(2)			
H$_2$Os$_3$Rh(CO)$_{10}$(acac)	1	Os	–	2.900(1)	–	X	394
H$_2$Os$_3$Ni(CO)$_{10}$(PPh$_3$)$_2$	1	Os	1.819^f	2.984(1)	110^f	X	394
	1	Os, Ni	1.758^f	2.733(1)	101^f		
H$_2$Os$_3$Pt(CO)$_{10}$(PPh$_3$)$_2$	1	Os	–	3.043(2)	–	X	393
	1	Os, Pt	–	2.848(2)	–		
H$_2$Os$_3$Pt(CO)$_{10}$(PCy$_3$)	1	Os, Pt	1.85^f	2.863(1)	–	X	376
	1	Os	1.85^f	2.789(1)			
H$_2$Os$_2$Pt$_2$(CO)$_8$(PPh$_3$)$_2$	1	Os, Pt	–	2.863(1)	–	X	376
H$_3$Os$_4$(CO)$_{12}$I	1	Os	–	3.010(4)	–	X	378
[H$_2$Os$_4$(CO)$_{12}$]$^{2-}$[(Ph$_3$P)$_2$N]$_2^+$	1	Os	–	2.934(2)	–	X	379
H$_2$Os$_4$Se$_2$(CO)$_{12}$	1	Os	–	2.965(1)	–	X	389
H$_3$Os$_4$(CO)$_{11}$(HC=CHPh)	1	Os	1.80(2)	2.787(3)	102(1)	N	202
	1	Os	1.81(2)	2.963(1)	110(1)	N	

Table 5 (continued)

Complex[a]	n[b]	M	M−H (Å)[c]	M−M (Å)[c]	M−H−M (°)[c]	Type[d]	Ref.
$H_3Os_4(CO)_{11}(C_6H_9)$	1	Os	−	2.918(1)	−	X	223
$H_2Os_5(CO)_{16}$	1	Os	−	2.962(5)	−	X	225
$[HOs_5(CO)_{15}]^-[(Ph_3P)_2N]^+$	1	Os	−	2.867	−	X	224
$HOs_5C(CO)_{14}[OP(OMe)_2]$	1	Os	−	2.914(2)	−	X	226b
$HOs_5C(CO)_{13}$- $[OP(OMe)OP(OMe)_2]$	1	Os	−	2.914(3)	−	X	226a
$H_2Os_3Re_2(CO)_{20}$	1	Os	−	3.069(1)	−	X	370
$H_2Os_6(CO)_{18}$	1	Os	−	2.928(4)	−	X	37, 241
$[HOs_3(CO)_{10}]_2(H_2CS_2)$	1	Os	−	−	−	X	383
$[Os_6(CO)_{17} \cdot CO_2 \cdot HOs_3(CO)_{10}]^-$- $[(Ph_3P)_2N]^+$	1	Os	−	2.895	−	X	380
$\{H_3Co_2[MeC(CH_2AsPh_2)_3]_2\}^+$- $[BPh_4]^-$	3	Co	1.70(4)	2.377(8)	88.7	X	187
$[Rh(C_5Me_5)Cl]_2(H)(Cl)$	1	Rh	1.85(5)	2.906(1)	104(4)	X	191
$H_2Rh_2[P(OPr^i)_3]_4$	2	Rh	1.808(12)	2.647(13)	94.1(4)	N	182a 182c
$H_3Rh_3[P(OCH_3)_3]_6$	1	Rh	1.765(11)	2.813(7)	105.7(6)	N	182a 182b
$[H_3Rh_3(C_5Me_5)_3O]^+[PF_6]^-$	1	Rh	1.7	2.761(1)	100	X	391
$[H_4Rh_4(C_5Me_5)_4]^{2+}[BF_4]_2^-$	1	Rh	−	2.829(1)	−	X	215
$HIr_2(C_5Me_5)_2Cl_3$	1	Ir	1.94(7)	2.903(1)	97(3)	X	192
$[H_5Ir_2(PPh_3)_4]^+[PF_6]^-$	3	Ir	−	2.518	−	X	188
$\{H_5Ir_2[Ph_2P(CH_2)_3PPh_2]_2\}^+$- $[BF_4]^-$	3	Ir	−	2.514(1)	−	X	81
$[H_3Ir_2(C_5Me_5)_2]^+[BF_4]^-$	3	Ir	1.75(3)	2.458(6)	89.5(8)	N	51, 382
$\{H_7Ir_3[Ph_2P(CH_2)_3PPh_2]_3\}^{2+}$- $[BF_4]_2^-$	1	Ir	−	2.772(1)	−	X	81
$[HNi_2(CO)_6]^-[(Ph_3P)_2N]^+$	1	Ni	1.70[f]	2.864(3)	−	X	34
$H_2Ni_2[Cy_2P(CH_2)_3PCy_2]_2$	2	Ni	1.58(4)	2.441(1)	101	X	181
$[HPt(SiEt_3)(PCy_3)]_2$	2	Pt	−	2.692(3)	−	X	159
$[H_2Pt_2(Ph)(PEt_3)_4]^+[BPh_4]^-$	1	Pt	−	3.09	−	X	176
$\{H_3Pt_2[Bu^t_2P(CH_2)_3PBu^t_2]_2\}^+$- $[BPh_4]^-$	1	Pt	−	2.768(2)	−	X	177
$H_6Cu_6(PPh_3)_6$	1	Cu	−	2.655(17)	−	X	228
$H_4Th_2(C_5Me_5)_4$	2	Th	2.29(3)	4.007(8)	122(4)	N	55

a   In this table we have adopted the convention (for ionic compounds) of listing the hydrido species first
b   1 = $M(\mu$-H)M, 2 = $M(\mu$-H)$_2$M, 3 = $M(\mu$-H)$_3$M, 4 = $M(\mu$-H)$_4$M
c   Whenever appropriate, averages are reported, Numbers in parentheses give the estimated standard deviation of the last reported digit. No estimated error is given in cases when hydrogen atom positions were varied in least-squares calculations, or when no standard deviation is reported
d   N = neutron, X = X-ray
e   Although at 22 °C the Cr−H−Cr bond in $[(Ph_3P)_2N]^+[HCr_2(CO)_{10}]^-$ appears linear, the large root mean square amplitudes of thermal displacement normal to the Cr−Cr vector [0.42(4) and 0.53(3) Å] strongly suggest the possibility of a fourfold or radially disordered bridging H atom (Ref. 171). In a recent study of $[(Ph_3P)_2N]^+[DCr_2(CO)_{10}]^-$ at 17 K, a fourfold disor-

dered model was refined, leading to Cr–D–Cr angles of approximately 156°, although it was not possible to resolve the disordered D atom positions in a difference-Fourier map (Ref. 172)

f   Distances and angles corresponding to calculated or estimated H positions

g   Values derived from nematic-phase proton nmr data, using heavy atom positions derived from an X-ray analysis (Ref. 40)

Table 6. Compounds with triply-bridging (face-bridging) H atoms

Compound[a]	M–H (Å)[b]	M–M (Å)[b]	M–H–M (°)[b]	Type[c]	Ref.
(A) Metal clusters					
$H_4Re_4(CO)_{12}$	1.77	2.913(8)	–	X	12
$[HFe_3Ni(CO)_{12}]^-[NMe_3Bz]^{+d}$	· 1.57(3)[d]	2.515(1)[d]	106(2)[d]	X	239
$H_2Ru_6(CO)_{18}$	–	2.954(3)	–	X	240
$[HOs_6(CO)_{18}]^-[(Ph_3P)_2N]^+$	–	2.973(3)	–	X	241
$HFeCo_3(CO)_9[P(OMe)_3]_3$	1.734(4)	2.489(3)	91.8(2)	X, N	237, 238
$H_4Co_4Cp_4$	1.67(7)	2.467(2)	–	X	230
HRh_3Cp_4	–	2.725(3)	–	X	229
$\{H_7Ir_3[Ph_2P(CH_2)_3PPh_2]_3\}^{2+}$-$[BF_4]_2^-$	–	2.772(1)	–	X	81
$[H_7Ir_3(py)_3(PCy_3)_3]^{2+}[PF_6]_2^-$	–	2.765(1)	–	X	80
$H_3Ni_4Cp_4$	1.691(8)	2.469(6)	93.9(3)	X, N	231, 232
(B) Metalloborane clusters					
$H_2Fe(MeC_2B_4H_4)_2$ [e]	–	–	–	X	244
$Cp_2Fe(H)Co(Me_2C_2B_3H_3)$ [f]	–	2.557(1)	–	X	243
$Co_2(B_4H_6)Cp_2$ [f]	1.55(3)	2.557(1)	111.1(3)	X	114
$Co_3(B_3H_5)Cp_3$ [f]	–	2.725(3)	–	X	229

a   In this table we have adopted the convention (for ionic compounds) of listing the hydrido species first

b   Whenever appropriate, averages are reported. Numbers in parentheses give the estimated standard deviation of the last reported digit. No estimated error is given in cases when hydrogen atom positions were not varied in least-squares calculations, or when no standard deviation is reported

c   N = neutron, X = X-ray

d   Face-Bridging H atom on Fe_2Ni face. Individual values: Ni–H = 1.55(3) Å; Fe–H = 1.56(2), 1.61(3) Å

e   Face-Bridging H atom on MB_2 face

f   Face-Bridging H atom on M_2B face

Table 7. Compounds with interstitial atoms[a]

Compound[b]	M–H (Å)[c]	M–M (Å)[c]	M–H–M (°)[c]	Type[d]	Ref.
HNb_6I_{11}	2.01	2.84	90.0	N[e]	248
$[HRu_6(CO)_{18}]^-[AsPh_4]^+$	2.04(1)	2.87(1)	90.0	X, N	249, 251
$[HCo_6(CO)_{15}]^-[(Ph_3P)_2N]^+$	1.824(13)	2.579(15)	90.0(6)	N	246
$[H_2Rh_{13}(CO)_{24}]^{3-}[PPh_3Bz]_3^+$	–	2.824(3)	–	X	255
$[H_2Rh_{13}(CO)_{24}]^{3-}[NBu_4^n]_3^+$	–	2.823(3)	–	X	256
$[H_3Rh_{13}(CO)_{24}]^{2-}[(Ph_3P)_2N]_2^+$	–	2.83(1)	–	X	254
$[HNi_{12}(CO)_{21}]^{3-}[AsPh_4]_3^+$ f	1.84	2.659	92.5	N	247
	2.00	2.439	75.1		
$[H_2Ni_{12}(CO)_{21}]^{2-}[PPh_4]_2^+$ f	1.72	2.682	102.5	N	247
	2.22	2.425	66.2		
$[HRh_{14}(CO)_{25}]^{3-}[NEt_4]_3^+$	–	2.757	–	X	392

a   All H atoms in this table have been located in octahedral cavities, except those of the Rh_{13} clusters, which are believed to exist in square pyramidal cavities

b   In this table we have adopted the convention (for ionic compounds) of listing the hydrido species first

c   Whenever appropriate, averages are reported. Numbers in parentheses give the estimated standard deviation of the last reported digit. No estimated error is given in cases when hydrogen atom positions were not varied in least-squares calculations, or when no standard deviation is reported

d   N = neutron, X = X-ray

e   The structure of HNb_6I_{11} was solved by neutron diffraction on a powdered sample

f   The H atoms in the Ni_{12} clusters are asymmetrically positioned within the octahedral cavities

Abbreviations Used in this Paper

acac	acetylacetonate $[CH_3C(O)CHC(O)CH_3]^-$
Bu^n	n-butyl $[CH_2CH_2CH_2CH_3]$
Bu^t	tert-butyl $[C(CH_3)_3]$
Bz	benzyl $[CH_2C_6H_5]$
Cp	cyclopentadienyl $[C_5H_5]$
Cy	cyclohexyl $[C_6H_{11}]$
dien	diethylenetriamine $[H_2NCH_2CH_2NHCH_2CH_2NH_2]$
DMF	dimethyl formamide $[HCON(CH_3)_2]$
Et	ethyl $[C_2H_5]$
Me	methyl $[CH_3]$
Me_2pz	dimethyl pyrazolyl $[N_2C_3H(CH_3)_2]$
Ph	phenyl $[C_6H_5]$
Pr^i	iso-propyl $[CH(CH_3)_2]$
Pr^n	n-propyl $[CH_2CH_2CH_3]$
py	pyridine $[C_5H_5N]$
pz	pyrazolyl $[N_2C_3H_3]$
THF	tetrahydrofuran $[C_4H_8O]$

References

1. B. A. Frenz, J. A. Ibers: in Transition Metal Hydrides (E. L. Muetterties, ed.) Marcel Dekker, Inc., New York (1971); pg. 33
2. A. P. Ginsberg: Trans. Met. Chem. *1*, 112, (1965)
3. H. D. Kaesz, R. B. Salliant: Chem. Rev. *72*, 231 (1972)
4. a) R. Bau and T. F. Koetzle, Pure Appl. Chem., *50*, 55 (1978)
   b) R. Bau, et al. Acc. Chem. Res. *12*, 176 (1979)
5. M. R. Churchill: Adv. Chem. Ser. *167*, 36 (1978)
6. J. A. Ibers: ibid. *167*, 27 (1978)
7. "Transition Metal Hydrides", (R. Bau, ed.) Amer. Chem. Soc. Adv. Chem. Ser., Vol. 167(1978)
8. a) W. M. Mueller, J. P. Blackledge, G. G. Libowitz: Metal Hydrides, Academic Press, New York (1968)
   b) K. M. MacKay, Hydrogen Compounds of the Metallic Elements, Barnes and Noble, New York (1966)
9. P. G. Owsten, J. M. Partridge, J. M. Rowe: Acta. Cryst. *13*, 246 (1960)
10. S. J. La Placa, W. C. Hamilton, J. A. Ibers: Inorg. Chem. *3*, 1491 (1964)
11. M. R. Churchill et al.: J. Am. Chem. Soc. *90*, 7135 (1968)
12. R. D. Wilson, R. Bau: ibid. *98*, 4687 (1976)
13. a) G. R. Wilkes: Ph. D. Thesis, Univers. Wisconsin, Madison, Wisconsin. 1965
    b) M. R. Churchill, J. P. Hutchinson: Inorg. Chem. *17*, 3528 (1978)
14. H. D. Kaesz et al.: J. Am. Chem. Soc. *91*, 1021 (1969)
15. R. Bau et al.: Inorg. Chem., *14*, 3021 (1975)
16. M. R. Churchill, B. G. DeBoer, F. J. Rotella: Inorg. Chem. *15*, 1843 (1976)
17. a) S. J. La Placa, J. A. Ibers: J. Am. Chem. Soc. *85*, 3501 (1963)
    b) S. J. La Placa, J. A. Ibers: Acta. Cryst. *18*, 511 (1965)
18. S. W. Kirtley, J. P. Olsen, R. Bau: J. Am. Chem. Soc. *98*, 4532 (1973)
19. E. Huber-Buser: Z. Krist. *133*, 150 (1971)
20. P. Dapporto et al.: ibid. *149*, 69 (1979)
21. a) J. L. Petersen, J. M. Williams: Inorg. Chem. *17*, 1308 (1978)
    b) J. L. Petersen, L. F. Dahl, J. M. Williams: J. Am. Chem. Soc. *96*, 6610 (1974)
    c) R. J. Doedens, L. F. Dahl: ibid. *87*, 2576 (1965)
22. G. E. Bacon: Neutron Diffraction, 3rd. ed., Clarendon Press, Oxford (1975)
23. a) J. P. Olsen et al.: J. Am. Chem. Soc. *96*, 6621 (1974)
    b) M. A. Andrews et al.: Chem. Commun. 181, (1973)
24. J. M. Carpenter et al.: Physics Today Dec. 1979, pg. 42
25. R. B. Calvert et al.: J. Am. Chem. Soc. *100*, 6240 (1978)
26. a) C. K. Johnson et al.: J. Am. Chem. Soc. *87*, 1802 (1965)
    b) M. D. Fronckowiak, R. K. McMullan: Amer. Cryst. Assoc., Program and Abstr., Eufala, Alabama, Abstr. N4, pg. 36 (1980)
27. A. Tippe, W. C. Hamilton: Inorg. Chem. *8*, 464 (1969)
28. P. L. Johnson et al.: J. Am. Chem. Soc. *100*, 2709 (1978)
29. F. Takusagawa, A. Fumagelli, T. F. Koetzle, S. G. Shore, T. A. Schmitkons, A. V. Fratini, K. W. Morse, C. Y. Wei, R. Bau: submitted for publication
30. a) A. G. Orpen et al.: J.C.S. Chem. Comm. 723 (1978)
    b) A. G. Orpen et al.: Acta. Cryst. *B34*, 2466 (1978)
31. W. C. Hamilton, J. A. Ibers: Acta. Cryst. *16*, 1209 (1963)
32. W. H. Baur: Acta. Cryst. *19*, 909 (1965)
33. V. G. Albano, P. L. Bellon: J. Organomet. Chem. *37*, 151 (1972)
34. G. Longoni, M. Manassero, M. Sansoni: J. Organomet. Chem. *174*, C41 (1979)
35. G. Ciani et al.: J.C.S. Dalton, 1943 (1976)
36. A. Bertolucci et al.: J. Organomet. Chem. *117*, C37 (1976)
37. a) A. G. Orpen: ibid. *159*, C1 (1978)
    b) A. G. Orpen: J. C. S. Dalton, submitted for publication

38. R. J. Klinger, J. C. Huffman, J. K. Kochi: J. Am. Chem. Soc. *102*, 208 (1980)
39. A. J. Schultz et al.: Inorg. Chem. *16*, 3303 (1977)
40. G. M. Sheldrick, J. P. Yesinowski: J.C.S. Dalton, 873 (1975)
41. G. M. Sheldrick: private communication
42. J. C. Green, D. M. P. Mingos, E. A. Seddon: J. Organomet. Chem. *185*, C20 (1980)
43. W. N. Lipscomb: Boron Hydrides, Benjamin 1963
44. S. C. Abrahams, A. P. Ginsberg, K. Knox: Inorg. Chem., *3*, 558 (1964)
45. S. J. La Placa et al.: ibid. *8*, 1928 (1969)
46. E. A. McNeill, F. R. Scholer: J. Am. Chem. Soc. *99*, 6243 (1977)
47. a) L. J. Guggenberger et al.: J. Am. Chem. Soc. *94*, 1135 (1972)
    b) P. Meakin et al.: J. Am. Chem. Soc. *92*, 3482 (1970)
48. M. B. Smith, R. Bau: J. Am. Chem. Soc. *95*, 2388 (1973)
49. G. Ciani et al.: J. Organomet. Chem. *152*, 85 (1978)
50. G. Ciani et al.: J. Organomet. Chem. *157*, 199 (1978)
51. R. Bau et al.: Adv. Chem. Ser. *167*, 73 (1978)
52. G. R. Clark, B. W. Skelton, T. N. Waters: Inorg. Chim. Acta. *12*, 235 (1975)
53. J. P. Jesson: in Transition Metal Hydrides (E. L. Muetterties, ed.) pg. 75, Marcel Dekker, New York (1971)
54. R. D. Wilson et al.: J. Am. Chem. Soc. *99*, 1775 (1977)
55. R. W. Broach et al.: Science *203*, 172 (1979)
56. a) M. Gerloch, R. Mason: J. Chem. Soc. 296 (1965)
    b) S. C. Abrahams, A. P. Ginsberg: Inorg. Chem. *5*, 500 (1966)
57. a) L. J. Guggenberger: Inorg. Chem. *12*, 294 (1973)
    b) L. J. Guggenberger, F. N. Tebbe: J. Am. Chem. Soc. *93*, 5924 (1971)
58. C. Couldwell, K. Prout: Acta. Cryst. *B35*, 335 (1979)
59. D. W. Hart, R. Bau, T. F. Koetzle: J. Am. Chem. Soc. *99*, 7557 (1977)
60. R. Bau et al.: J. Am. Chem. Soc. *99*, 3872 (1977)
61. B. A. Frenz, J. A. Ibers: Inorg. Chem. *9*, 2403 (1970)
62. R. W. Baker, P. Pauling: Chem. Commun. 1495 (1969)
63. R. W. Baker et al.: Chem. Commun. 1077 (1970)
64. a) A. Orlandini, L. Sacconi: Cryst. Struct. Commun. *4*, 107 (1975)
    b) A. Orlandini, L. Sacconi: Inorg. Chim. Acta. *19*, 61 (1976)
65. a) C. A. Ghilardi, S. Midollini, L. Sacconi: Cryst. Struct. Commun. *4*, 149 (1975)
    b) C. A. Ghilardi, S. Midollini, L. Sacconi: Inorg. Chem. *14*, 1790 (1975)
66. A. Orlandini, L. Sacconi: Cryst. Struct. Commun. *4*, 157 (1975)
67. a) L. Sacconi et al.: Inorg. Chem. *14*, 1380 (1975)
    b) C. A. Ghilardi, L. Sacconi: Cryst. Struct. Commun. *3*, 415 (1974)
68. P. Stoppioni, P. Dapporto: Cryst. Struct. Commun. *8*, 15 (1979)
69. L. Sacconi, A. Orlandini, S. Midollini: Inorg. Chem. *13*. 2850 (1974)
70. a) R. Ros et al.: Inorg. Chim. Acta. *29*, L187 (1978)
    b) A. del Pra et al.: J. Chem. Soc. Dalton 1862 (1979)
71. E. O. Fischer, A. Frank: Chem. Ber. *111*, 3740 (1978)
72. M. F. Hawthorne, C. B. Knobler, M. Delaney: submitted for publication
73. a) T. H. Tulip, J. A. Ibers: J. Am. Chem. Soc. *100*, 3252 (1978)
    b) T. H. Tulip, J. A. Ibers: J. Am. Chem. Soc. *101*, 4201 (1979)
74. T. V. Ashworth, E. Singleton, M. Laing: J. Organometal. Chem. *117*, C113 (1976)
75. J. C. McConway et al.: J.C.S. Chem. Comm. 327 (1974)
76. a) F. A. Cotton, D. L. Hunter, B. A. Frenz: Inorg. Chim. Acta. *15*, 155 (1975)
    b) F. A. Cotton, B. A. Frenz, D. L. Hunter: Chem. Comm. 755 (1974)
77. S. Komiya et al.: J. Am. Chem. Soc. *98*, 3874 (1976)
78. a) J. R. Shapley et al.: ibid. *97*, 4145 (1975)
    b) M. R. Churchill, B. G. DeBoer: Inorg. Chem. *16*, 878 (1977)
79. G. Ciani, V. G. Albano, A. Immirzi: J. Organomet. Chem. *121*, 237 (1976)
80. D. F. Chodosh et al.: J. Organomet. Chem. *161*, C67 (1978)
81. H. H. Wang, L. H. Pignolet: Inorg. Chem. *19*, 1470 (1980)

82. R. O. Moyer, Jn., R. Lindsay, D. N. Marks: Adv. Chem. Ser. *167*, 366 (1978)
83. R. O. Moyer, Jn. et al.: J. Solid State Chem. *3*, 541 (1971)
84. L. B. Lundberg, D. T. Cromer, C. B. Magee: Inorg. Chem. *11*, 400 (1972)
85. T. J. Marks, J. R. Kolb: Chem. Rev. *77*, 263 (1977)
86. a) E. B. Lobkovskii, S. E. Kravchenko, K. N. Semenenko: Zhur. Strukt. Khim. *18*, 389 (1977); J. Struct. Chem. *18*, 312 (1972)
    b) B. G. Segal, S. J. Lippard: Inorg. Chem. *17*, 844 (1978)
87. R. R. Rietz et al.: Inorg. Chem. *17*, 658 (1978)
88. A. Zalkin et al.: ibid. *17*, 661 (1978)
89. R. R. Rietz et al.: ibid. *17*, 653 (1978)
90. a) J. L. Atwood et al.: J.C.S. Chem. Commun. 593 (1977)
    b) C. Kutal et al.: Inorg. Chem. *17*, 3558 (1978)
91. K. M. Melmed et al.: J. Am. Chem. Soc. *96*, 69 (1974)
92. S. W. Kirtley et al.: J. Am. Chem. Soc. *99*, 7154 (1977)
93. a) M. Nakajima et al.: J.C.S. Dalton. 385 (1977)
    b) M. Nakajima et al.: Chem. Comm. 80 (1975)
94. T. Saito et al.: J.C.S. Dalton, 482 (1978)
95. P. Dapporto et al.: Inorg. Chem. *15*, 2768 (1976)
96. a) K. M. Melmed, D. Coucouvanis, S. J. Lippard: Inorg. Chem. *12*, 232 (1973)
    b) E. I. Mamaeva, I. Hargittai, A. P. Spiridonov: Inorg. Chim. Acta. *25*, L123 (1977)
97. N. I. Kirillova, A. I. Gusev, Y. T. Struchkov: J. Struct. Chem. *15*, 622 (1974); Zhur. Strukt. Khim. *15*, 718 (1974)
98. E. J. Corey, N. J. Cooper, W. M. Canning, W. N. Lipscomb, T. K. Koetzle: Inorg. Chem., to be published
99. P. H. Bird, M. R. Churchill: Chem. Commun. 403, (1967)
100. a) V. Plato, K. Hedberg: Inorg. Chem. *10*, 590 (1971)
    b) V. P. Spiridonov, G. I. Mamaeva: J. Struct. Chem. *10*, 120 (1969)
101. a) E. R. Bernstein, W. C. Hamilton, T. A. Keiderling, W. J. Kennelly, S. J. La Placa, S. J. Lippard, T. J. Marks, J. J. Mayerle: unpublished results, reported in the Ph. D. dissertation of T. A. Keiderling, Princeton University, pages 174–188 (1974)
    b) T. J. Marks, J. M. Williams: private communication
102. E. R. Bernstein et al.: Inorg. Chem. *11*, 3009 (1972)
103. S. J. Lippard, K. M. Melmed: Inorg. Chem. *8*, 2755 (1969)
104. L. J. Guggenberger: Inorg. Chem. *9*, 367 (1970)
105. S. J. Hildebrandt, D. F. Gaines, J. C. Calabrese: Inorg. Chem. *17*, 790 (1978)
106. J. C. Calabrese et al.: J. Am. Chem. Soc. *96*, 6318 (1974)
107. a) G. B. Dunks, M. F. Hawthorne: Acc. Chem. Res. *6*, 124 (1973)
    b) R. N. Grimes: ibid. *11*, 420 (1978)
108. a) J. W. Lott et al.: J. Am. Chem. Soc. *95*, 3042 (1973)
    b) J. W. Lott, D. F. Gaines: Inorg. Chem. *13*, 2262 (1974)
109. J. R. Pipal, R. N. Grimes: ibid. *16*, 3251 (1977)
110. T. P. Fehlner et al.: J. Am. Chem. Soc. *98*, 7085 (1976)
111. N. N. Greenwood et al.: J. Chem. Soc. Dalton 117, (1979)
112. L. G. Sneddon, D. Voet: J.C.S. Chem. Commun. 118 (1976)
113. M. Mangion et al.: J. Am. Chem. Soc. *101*, 754 (1979)
114. J. R. Pipal, R. N. Grimes: Inorg. Chem. *18*, 252 (1979)
115. J. T. Gill, S. J. Lippard: Inorg. Chem. *14*, 751 (1975)
116. T. E. Paxson et al.: Inorg. Chem. *13*, 2772 (1974)
117. R. A. Love, R. Bau: J. Am. Chem. Soc. *94*, 8274 (1972)
118. G. Allegra et al.: Cryst. Struct. Commun. *3*, 69 (1974)
119. R. T. Baker et al.: J. Am. Chem. Soc. *100*, 8266 (1978)
120. H. M. Colquhoun, T. J. Greenhough, M. G. H. Wallbridge: J. C. S. Chem. Comm. 192 (1980)
121. H. D. Kaesz et al.: J. Am. Chem. Soc. *87*, 2753 (1965)
122. C. A. Kosky, P. Ganis, G. Avitabile: Acta. Cryst. *B27*, 1859 (1971)
123. F. A. Cotton, M. Jeremic, A. Shaver: Inorg. Chim. Acta. *6*, 543 (1972)

124. L. J. Guggenberger, F. N. Tebbe: J. Am. Chem. Soc. *95*, 7870 (1973)
125. F. N. Tebbe, L. J. Guggenberger: J.C.S. Chem. Commun. 227 (1973)
126. a) R. A. Forder, K. Prout: Acta. Cryst. *B30* 2312 (1974)
    b) S. J. Rettig et al.: Acta. Cryst. *B30*, 666 (1974)
127. T. J. McNeese, S. S. Wreford, B. M. Foxman: J.C.S. Chem. Commun. 500 (1978)
128. G. M. Parshall: Acc. Chem. Res. *8*, 113 (1975)
129. D. M. Roe et al.: J.C.S. Chem. Commun. 1273 (1972)
130. F. A. Cotton, T. LaCour, A. G. Stanislowski: J. Am. Chem. Soc. *96*, 754 (1974)
131. F. A. Cotton, V. W. Day: J.C.S. Chem. Commun. 415 (1974)
132. a) S. Trofimenko: J. Am. Chem. Soc. *90*, 4754 (1968)
    b) S. Trofimenko: Inorg. Chem. *9*, 2493 (1970)
133. a) J. M. Williams et al.: J. Am. Chem. Soc. *100*, 7407 (1978)
    b) R. K. Brown et al.: ibid. *102*, 981 (1980)
134. S. J. La Placa, J. A. Ibers: Inorg. Chem. *4*, 778 (1965)
135. N. A. Bailey et al.: Chem. Commun. 237 (1965)
136. A. C. Skapski, P. G. H. Troughton: Chem. Commun. 1230 (1968)
137. K. W. Muir, J. A. Ibers: Inorg. Chem. *9*, 440 (1970)
138. M. J. Bennett et al.: Inorg. Chim. Acta. *12*, L9 (1975)
139. S. Otsuka et al.: J. Am. Chem. Soc. *98*, 5850 (1976)
140. Y. W. Yared et al.: J. Am. Chem. Soc. *99*, 7076 (1977)
141. G. K. Anderson et al.: Inorg. Chim. Acta. *29*, L193 (1978)
142. R. A. Andersen, R. A. Jones, G. Wilkinson: J. Chem. Soc. Dalton 446 (1978)
143. P. B. Hitchcock et al.: ibid. 1929 (1979)
144. J. Kopf, H. J. Vollmer, W. Kaminsky: Cryst. Struct. Comm. *9*, 271 (1980)
145. L. G. Kuz'mina, Yu. T. Struchkov: ibid. *8*, 715 (1979)
146. B. Granoff, R. A. Jacobson: Inorg. Chem. *7*, 2328 (1965)
147. M. Pasquali et al.: J. Am. Chem. Soc. *100*, 4918 (1978)
148. A. J. Schultz et al.: J. Am. Chem. Soc. *101*, 1593 (1979)
149. M. R. Churchill, W. J. Youngs: Inorg. Chem. *18*, 1930 (1974)
150. M. R. Churchill, W. J. Youngs: ibid. *18*, 2454 (1979)
151. a) M. A. Beno et al.: J. Am. Chem. Soc. *102*, 4542 (1980)
    b) J. M. Williams et al., submitted for publication
152. a) M. Manassero, M. Sansoni, G. Longoni: J.C.S. Chem. Comm. 919 (1976)
    b) M. Manassero, M. Sansoni, G. Longoni: private communication.
153. R. B. Calvert, J. R. Shapley: J. Am. Chem. Soc. *100*, 7726 (1978)
154. M. Cowie, M. J. Bennett: Inorg. Chem. *16*, 2321 (1977)
155. M. Elder: ibid. *9*, 762 (1970)
156. M. J. Bennett, K. A. Simpson: J. Am. Chem. Soc. *93*, 7156 (1971)
157. M. Cowie, M. J. Bennett: Inorg. Chem. *16*, 2325 (1977)
158. a) M. Auburn et al.: J.C.S. Dalton 659 (1980)
    b) M. Ciriano et al.: Adv. Chem. Ser. *167*, 111 (1978)
159. a) M. Ciriano et al.: J.C.S. Dalton 801 (1978)
    b) M. Green et al.: J.C.S. Chem. Comm. 671 (1976)
160. J. A. K. Howard: private communication.
161. R. A. Smith, M. J. Bennett: Acta. Cryst. *B33*, 1113 (1977)
162. Chem. Eng. News June 8, pg. 75 (1970)
163. R. A. Smith, M. J. Bennett: Acta. Cryst., *B33*, 1118 (1977)
164. D. E. Crotty et al.: Inorg. Chem. *16*, 2346 (1977)
165. a) P. T. Mosely, H. M. M. Shearer, C. B. Spencer: Acta. Cryst. *A25*, S169 (1969)
    b) N. A. Bell et al.: J.C.S. Chem. Comm. 359 (1980)
166. F. W. J. Benfield et al.: ibid. 759 (1973)
167. R. A. Forder, K. Prout: Acta. Cryst. *B30*, 2318 (1974)
168. R. A. Love et al.: J. Am. Chem. Soc. *98*, 4491 (1976)
169. a) L. B. Handy et al.: ibid. *88*, 366 (1966)
    b) L. B. Handy, J. K. Ruff, L. F. Dahl: J. Am. Chem. Soc. *92*, 7312 (1970)

170. J. Roziere et al.: J. Am. Chem. Soc. *99*, 4497 (1977)
171. J. L. Petersen et al.: Inorg. Chem. *17*, 3460 (1978)
172. J. L. Petersen et al.: ibid. *18*, 3493 (1979)
173. R. D. Wilson, S. A. Graham, R. Bau: J. Organomet. Chem. *91*, C49 (1975)
174. D. W. Hart, R. Bau, T. F. Koetzle: to be published
175. a) M. Y. Darensbourg et al.: J. Am. Chem. Soc. *101*, 2631 (1979)
     b) J. L. Petersen, A. Masino, R. P. Stewart, Jr.: J. Organomet. Chem. submitted for publication
176. G. Bracher et al.: Angew. Chem. *90*, 826 (1978) Internat. Edit. *17*, 778 (1978)
177. T. H. Tulip et al.: Inorg. Chem. *18*, 2239 (1979)
178. a) J. A. Labinger, K. S. Wong, W. R. Scheidt: J. Am. Chem. Soc. *100*, 3254 (1978)
     b) K. S. Wong, W. R. Scheidt, J. A. Labinger: Inorg. Chem. *18*, 136 (1979)
179. M. J. Bennett et al.: J. Am. Chem. Soc. *94*, 6232 (1972)
180. a) M. R. Churchill, S. W. Y. Chang: Inorg. Chem. *13*, 2413 (1974)
     b) M. R. Churchill et al.: Chem. Commun. 691 (1973)
181. B. L. Barnett et al.: Chem. Ber. *110*, 3900 (1977)
182. a) R. K. Brown et al.: Proc. Nat. Acad. Sci. USA *76*, 2099 (1979)
     b) R. K. Brown et al.: Inorg. Chem. *19*, 370 (1980)
     c) R. G. Teller, T. F. Koetzle, J. M. Williams, A. J. Sivak and E. L. Muetterties, Inorg. Chem.,
        in press
183. N. W. Alcock et al.: Chem. Commun. 1160 (1979)
184. R. W. Broach, J. M. Williams: Inorg. Chem. *18*, 314 (1979)
185. a) A. Bertolucci et al.: J. Organomet. Chem. *113*, C61 (1976)
     b) G. Ciani, A. Sironi, V. G. Albano: J.C.S. Dalton 1667 (1977)
186. G. Ciani et al.: J. Organomet. Chem. *136*, C49 (1977)
187. a) P. Dapporto, S. Midollini, L. Sacconi: Inorg. Chem. *14*, 1643 (1975)
     b) P. Dapporto et al.: J. Am. Chem. Soc. *95*, 2021 (1973)
188. a) R. H. Crabtree et al.: J. Organomet. Chem. *113*, C7 (1976)
     b) R. H. Crabtree, H. Felkin, G. E. Morris: ibid. *141*, 205 (1977)
189. A. Dedieu, T. A. Albright, R. Hoffmann: J. Am. Chem. Soc. *101*, 3141 (1979)
190. a) R. Mason: Special Lectures 23rd Internat. Congr. Pure & Appl. Chem. *6*, 31
        (1971)
     b) R. Mason, D. M. P. Mingos: J. Organomet. Chem. *50*, 53 (1973)
191. M. R. Churchill, S. W. Y. Ni: J. Am. Chem. Soc. *95*, 2150 (1973)
192. M. R. Churchill, S. A. Julis: Inorg. Chem. *16*, 1488 (1977)
193. M. R. Churchill, R. A. Lashewycz: ibid. *18*, 1926 (1979)
194. M. R. Churchill, R. A. Lashewycz: Inorg. Chem. *18*, 3261 (1979)
195. F. A. Cotton, B. J. Kalbacher: ibid. *15*, 522 (1976)
196. A. Bino, F. A. Cotton: Angew. Chem. *91*, 356 (1979); Internat. Edit. *18*, 332 (1979)
197. A. Bino, F. A. Cotton: J. Am. Chem. Soc. *101*, 4150 (1979)
198. P. Chini, G. Longoni, V. G. Albano: Adv. Organomet. Chem. *14*, 285 (1976)
199. K. Wade: Adv. Inorg. Chem. Radiochem. *18*, 1 (1976)
200. E. L. Muetterties: Science *196*, 839 (1977)
201. A. L. Robinson: Science *194*, 1150 (1976); *194*, 1261 (1976)
202. A. G. Orpen: Ph. D. Thesis, Univers. Cambridge (1979)
203. V. F. Allen, R. Mason, P. B. Hitchcock: J. Organomet. Chem. *140*, 297 (1977)
204. M. R. Churchill, F. J. Hollander, J. P. Hutchinson: Inorg. Chem. *16*, 2697 (1977)
205. R. D. Adams, N. M. Golembski: Inorg. Chem. *18*, 1909 (1979)
206. R. E. Benfield et al.: Acta. Cryst. *B35*, 2210 (1979)
207. B. F. G. Johnson et al.: J. Chem. Soc. Dalton 616 (1979)
208. A. J. Schultz et al.: Inorg. Chem. *18*, 319 (1979)
209. M. R. Churchill, B. G. DeBoer: Inorg. Chem. *16*, 2397 (1977)
210. R. D. Wilson et al.: Inorg. Chem. *17*, 1271 (1978)
211. K. Sasvari et al.: Acta. Cryst. *B35*, 87 (1979)
212. a) J. R. Shapley et al.: J. Am. Chem. Soc. *99*, 7384 (1977)
     b) M. R. Churchill, R. A. Lashewycz: Inorg. Chem. *17*, 1950 (1978)

213. P. F. Jackson et al.: J.C.S. Chem. Commun. 920 (1978)
214. S. A. R. Knox, H. D. Kaesz: J. Am. Chem. Soc. *93*, 4594 (1971)
215. P. Espinet et al.: Inorg. Chem. *18*, 2706 (1979)
216. D. B. W. Yawney, R. J. Doedens: Inorg. Chem. *11*, 838 (1972)
217. a) C. J. Gilmore, P. Woodward: Chem. Commun. 1463 (1970)
     b) C. J. Gilmore, P. Woodward: J. Chem. Soc. (A) 3453 (1971)
218. G. Ciani, A. Sironi, V. G. Albano: J. Organomet. Chem. *136*, 339 (1977)
219. M. A. Andrews et al.: J. Am. Chem. Soc. *101*, 7245 (1979)
220. M. R. Churchill, F. J. Hollander: Inorg. Chem. *18*, 161 (1979)
221. M. R. Churchill, F. J. Hollander: Inorg. Chem. *18*, 843 (1979)
222. F. Takusagawa, A. Fumagalli, T. F. Koetzle, G. L. Geoffroy, W. L. Gladfelter, M. A. Bruck, R. Bau: manuscript in preparation
223. S. Bhaduri et al.: J. Chem. Soc. Dalton 562 (1979)
224. a) C. R. Eady et al.: J.C.S. Chem. Comm. 807 (1976)
     b) J. J. Guy, G. M. Sheldrick: Acta. Cryst. *B34*, 1722 (1978)
225. J. J. Guy, G. M. Sheldrick: ibid. *B34*, 1725 (1978)
226. a) A. G. Orpen, G. M. Sheldrick: ibid. *B34*, 1992 (1978)
     b) J. M. Fernandez et al.: Acta. Cryst. *B34*, 1994 (1978)
227. M. Tachikawa et al.: J. Am. Chem. Soc., *102*, 6648 (1980)
228. M. R. Churchill et al.: Inorg. Chem. *11*, 1818 (1972)
229. O. S. Mills, E. F. Paulus: J. Organomet. Chem. *11*, 587 (1968)
230. G. Huttner, H. Lorenz: Chem. Ber. *108*, 973 (1975)
231. a) G. Huttner, H. Lorenz: ibid. *107*, 996 (1974)
     b) J. Müller et al.: Angew. Chem. Internat. Edit. *12*, 1005 (1973); Angew. Chem. *85*, 1115 (1973)
232. T. F. Koetzle et al.: J. Am. Chem. Soc. *101*, 5631 (1979)
233. R. Hoffmann et al.: ibid. *100*, 6088 (1978)
234. P. Chini, L. Colli, M. Peraldo: Gazz. Chim. Ital. *90*, 1005 (1960)
235. J. W. White, C. J. Wright: J. Chem. Soc. A. 2843 (1971)
236. M. J. Mays, R. N. F. Simpson: J. Chem. Soc. A. 1444 (1968)
237. a) B. T. Huie, C. B. Knobler, H. D. Kaesz: J.C.S. Chem. Commun. 684 (1975)
     b) B. T. Huie, C. B. Knobler, H. D. Kaesz: J. Am. Chem. Soc. *100*, 3059 (1978)
238. R. G. Teller et al.: J. Am. Chem. Soc. *100*, 3071 (1978)
239. A. Fumagalli et al.: Amer. Cryst. Assoc. Program and Abstr., Eufala, Alabama, Abstr. D2, pg. 16 (1980)
240. a) M. R. Churchill, J. Wormald: J. Am. Chem. Soc. *93*, 5670 (1971)
     b) M. R. Churchill et al.: Chem. Comm. 458 (1970)
241. M. McPartlin et al.: J.C.S. Chem. Commun. 883 (1976)
242. J. R. Pipal, R. N. Grimes: Inorg. Chem. *16*, 3255 (1977)
243. R. N. Grimes, E. Sinn, R. B. Maynard: Inorg. Chem., *19*, 2384 (1980)
244. J. R. Pipal, R. N. Grimes: Inorg. Chem. *18*, 263 (1979)
245. K. Christmann et al.: J. Chem. Phys. *70*, 4168 (1979)
246. D. W. Hart et al.: Angew. Chem. Internat. Edit. *18*, 80 (1979); Angew. Chem. *91*, 81 (1979)
247. R. W. Broach et al.: Adv. Chem. Ser. *167*, 93 (1978)
248. A. Simon: Z. Anorg. Allg. Chem. *355*, 311 (1967)
249. C. R. Eady et al.: J.C.S. Chem. Comm. 945 (1976)
250. P. Chini et al.: Adv. Chem. Ser. *167*, 1 (1978)
251. P. F. Jackson et al.: J.C.S. Chem. Comm. 295 (1980)
252. R. D. Kelley, J. J. Rush, T. E. Madey: Chem. Phys. Lett. *66*, 159 (1979)
253. T. H. Upton, W. A. Goddard, III: Phys. Rev. Lett. *42*, 472 (1979)
254. V. G. Albano et al.: J.C.S. Chem. Comm. 859 (1975)
255. V. G. Albano et al.: J.C.S. Dalton 978 (1979)
256. R. Bau, C. Y. Wei, D. M. Ho, P. Chini, S. Martinengo, S. Campanella: unpublished
257. G. Ciani, A. Sironi, S. Martinengo: results to be published.

258. a) F. J. Hollander, D. Coucouvanis: J. Am. Chem. Soc. *99*, 6268 (1977)
     b) L. D. Lower, L. F. Dahl: J. Am. Chem. Soc. *98*, 5046 (1976)
259. N. I. Kirillova, A. I. Gusev, Y. T. Struchkov: J. Struct. Chem. *13*, 441 (1972); Zhur. Strukt. Khim. *13*, 473 (1972)
260. a) Yu. T. Struchkov: Pure Appl. Chem. *52*, 741 (1980)
     b) D. A. Lemenovskii et al.: J. Organomet. Chem. submitted for publication.
261. N. I. Kirillova et al.: J. Struct. Chem. 261 (1974); Zhur. Strukt. Khim. *15*, 288 (1974)
262. P. Meakin et al.: Inorg. Chem. *13*, 1025 (1974)
263. S. G. Davies et al.: J.C.S. Chem. Commun. 135 (1977)
264. a) M. L. H. Green et al.: ibid. 839 (1974)
     b) K. Prout, R. A. Forder: Acta. Cryst. *B31*, 852 (1975)
265. J. W. Byrne et al.: J.C.S. Chem. Commun. 662 (1977)
266. a) P. Meakin et al.: J. Am. Chem. Soc. *95*, 1467 (1973)
     b) L. J. Guggenberger: Inorg. Chem. *12*, 2295 (1973)
267. R. A. Jones et al.: J.C.S. Chem. Comm. 408 (1980)
268. a) D. Gregson et al.: ibid. 572 (1980)
     b) J. A. K. Howard: private communication
269. V. G. Albano et al.: J. Organomet. Chem. *34*, 353 (1972)
270. A. P. Ginsberg, S. C. Abrahams, P. B. Jamieson: J. Am. Chem. Soc. *95*, 4751 (1973)
271. R. G. Teller: Ph. D. Dissertation, Univers. Southern California, Chapter 2, 1978
272. M. R. Churchill, F. J. Hollander: Inorg. Chem. *16*, 2493 (1977)
273. K. Mertis et al.: J.C.S. Chem. Comm. 654 (1980)
274. L. Manojlovic-Muir, K. W. Muir, J. A. Ibers: Inorg. Chem. *9*, 447 (1970)
275. L. J. Guggenberger: Inorg. Chem. *12*, 1317 (1973)
276. G. K. Barker et al.: J.C.S. Chem. Commun. 169 (1978)
277. C. G. Pierpoint, R. Eisenberg: Inorg. Chem. *11*, 1094 (1972)
278. E. H. S. Wong, M. F. Hawthorne: J.C.S. Chem. Commun. 257 (1976)
279. A. C. Skapski, P. G. H. Troughton: Chem. Commun. 1230 (1968)
280. a) S. D. Ibekwe et al.: ibid. 433 (1969)
     b) U. A. Gregory et al.: J. Chem. Soc. (A) 1118 (1971)
281. L. D. Brown, J. A. Ibers: Inorg. Chem. *15*, 2788 (1976)
282. C. Potvin et al.: J. Organometal. Chem. *146*, 57 (1978)
283. A. Immirzi, A. Luccarelli: Cryst. Struct. Commun. *1*, 317 (1972)
284. K. D. Schramm, J. A. Ibers: Inorg. Chem. *16*, 3287 (1977)
285. a) L. D. Brown, J. A. Ibers: ibid. *15*, 2788 (1976)
     b) L. D. Brown and J. A. Ibers: J. Am. Chem. Soc. *98*, 1597 (1976)
286. L. D. Brown et al.: Inorg. Chem. *16*, 2728 (1977)
287. T. V. Ashworth et al.: J.C.S. Dalton 1816 (1977)
288. T. V. Ashworth, M. J. Nolte, E. Singleton: J. Organomet. Chem. *139*, C73 (1977); J.C.S. Dalton 1040 (1978)
289. a) A. C. Skapski, F. A. Stephens: J.C.S. Dalton, 390 (1974)
     b) A. C. Skapski, F. A. Stephens: Chem. Commun. 1008 (1969)
290. a) I. S. Kolomnikov et al.: J. Organomet. Chem. *59*, 349 (1973)
     b) I. S. Kolomnikov, A. I. Gusev, Y. T. Struchkov: J. Struct. Chem. 685 (1973)
291. R. A. Jones et al.: J.C.S. Dalton submitted for publication
292. M. Cowie, B. L. Haymore, J. A. Ibers: Inorg. Chem. *14*, 2617 (1975)
293. P. L. Orioli, L. Vaska: Proc. Chem. Soc. 333 (1962)
294. R. Mason, private communication. Mentioned briefly in L. Aslanov et al.: Chem. Comm. 30 (1970)
295. Z. Dawoodi, M. J. Mays, P. R. Raithby: J.C.S. Chem. Commun. 721 (1979)
296. J. M. Whitfield et al.: J.C.S. Dalton 407 (1977)
297. D. D. Titus et al.: Chem. Commun. 322 (1971)
298. B. R. Davis, N. C. Payne, J. A. Ibers: Inorg. Chem. *8*, 2719 (1969)
299. P. R. Hoffman et al.: Inorg. Chem. *15*, 2462 (1976)
300. G. E. Hardy et al.: Acta. Cryst. *B32*, 264 (1976)

301. S. H. Strauss et al.: Inorg. Chem. *17*, 3064 (1978)
302. K. W. Muir, J. A. Ibers: Inorg. Chem. *9*, 440 (1970)
303. T. Yoshida et al.: Inorg. Chim. Acta. *29*, L257 (1978)
304. B. A. Coyle, J. A. Ibers: Inorg. Chem. *11*, 1105 (1972)
305. M. Cowie, S. K. Dwight: ibid. *18*, 1209 (1979)
306. C. Crocker et al.: J.C.S. Chem. Commun. 498 (1979)
307. M. Ciechanowicz, A. C. Skapski, P. G. H. Troughton: Acta. Cryst. *B32*, 1673 (1976)
308. a) D. M. P. Mingos, J. A. Ibers: Inorg. Chem. *10*, 1479 (1971)
     b) G. R. Clark, J. M. Waters, K. R. Whittle: ibid. *13*, 1628 (1974)
309. G. del Piero, G. Perego, M. Cesari: Gazz. Chim. Ital. *105*, 529 (1975)
310. K. W. Muir, J. A. Ibers: J. Organomet. Chem. *18*, 175 (1969)
311. F. Glocking, M. D. Wilbey: J. Chem. Soc. (A) 1675, (1970)
312. P. L. Bellon et al.: J. Organomet. Chem. *157*, 209 (1978)
313. P. Bird, J. F. Harrod, K. A. Than: J. Am. Chem. Soc. *96*, 1222 (1974)
314. a) P. J. Roberts, G. Ferguson, C. V. Senoff: J. Organomet. Chem. *94*, C26 (1975)
     b) P. J. Roberts, G. Ferguson: Acta. Cryst. *B32*, 1513 (1976)
315. U. Behrens, L. Dahlenberg: J. Organomet. Chem. *116*, 103 (1976)
316. G. del Piero et al.: Cryst. Struct. Comm. *3*, 725 (1974)
317. P. L. Bellon et al.: J. Organomet. Chem. *157*, 209 (1978)
318. T. Debaerdemaeker: Cryst. Struct. Commun. *6*, 11 (1977)
319. N. N. Greenwood et al.: J.C.S. Dalton 1339 (1979)
320. M. D. Curtis, J. Greene, W. M. Butler: J. Organomet. Chem. *164*, 371 (1979)
321. G. R. Clark et al.: J. Organomet. Chem. *166*, 109 (1979)
322. J. J. Bonnet et al.: J. Am. Chem. Soc. *101*, 5940 (1979)
323. M. Angoletta et al.: J. Organomet. Chem. *81*, C40 (1974)
324. G. Ciani et al.: ibid. *150*, C17 (1978)
325. M. L. Schneider, H. M. M. Shearer: J.C.S. Dalton 354 (1973)
326. a) D. M. Roundhill: Adv. Chem. Ser. *167*, 160 (1978)
     b) R. E. Caputo et al.: Acta. Cryst. *B33*, 215 (1977)
327. P. G. Owsten, J. M. Partridge, J. M. Rowe: Acta. Cryst. *13*, 246 (1960)
328. R. Eisenberg, J. A. Ibers: Inorg. Chem. *4*, 773 (1965)
329. A. R. Kane, L. J. Guggenberger, E. L. Muetterties: J. Am. Chem. Soc. *92*, 2571 (1970)
330. A. Albinati et al.: Inorg. Chim. Acta. *18*, 219 (1976)
331. A. Immirzi, G. Bombieri, L. Toniolo: J. Organomet. Chem. *118*, 355 (1976)
332. E. A. V. Ebsworth et al.: J.C.S. Dalton 1167 (1978)
333. A. Immirzi et al.: Inorg. Chim. Acta. *12*, L23 (1975)
334. G. K. Barker et al.: J.C.S. Dalton 1687 (1979)
335. M. P. Brown et al.: J.C.S. Chem. Comm. 931 (1979)
336. T. Greiser, U. Puttfarcken, D. Rehder: Trans. Met. Chem. *4*, 168 (1979)
337. K. N. Semenenko, E. B. Lobkovskii, A. I. Shumakov: Zhur. Strukt. Khim. *17*, 1073
     (1976); J. Struct. Chem. *17*, 912 (1976)
338. G. Fachinetti et al.: J.C.S. Chem. Comm. 300 (1976)
339. D. F. Gaines, J. W. Lott, J. C. Calabrese: Inorg. Chem. *13*, 2419 (1974)
340. K. P. Callahan et al.: ibid. *13*, 2842 (1974)
341. a) S. J. Lippard, K. M. Melmed: ibid. *6*, 2223 (1967)
     b) S. J. Lippard, K. M. Melmed: J. Am. Chem. Soc. *89*, 3929 (1967)
     c) See also footnote 27 of J. T. Gill, S. J. Lippard: Inorg. Chem. *14*, 751 (1975)
342. A. J. Schultz, R. K. Brown, J. M. Williams, R. R. Schrock: submitted for publication
343. G. P. Pez et al.: J. Am. Chem. Soc. *101*, 6933 (1979)
344. A. D. Foust, W. A. G. Graham, R. P. Stewart, Jn.: J. Organomet. Chem. *54*, C22 (1973)
345. N. J. Cooper et al.: J.C.S. Chem. Comm. 145 (1977)
346. M. Akiyama et al.: J. Am. Chem. Soc. *101*, 2504 (1979)
347. R. J. Doedens, W. T. Robinson, J. A. Ibers: ibid. *89*, 4323 (1967)
348. a) H. D. Kaesz, R. Bau, M. R. Churchill: ibid. *89*, 2775 (1967)
     b) M. R. Churchill, R. Bau: Inorg. Chem. *6*, 2086 (1967)

349. B. T. Huie: Ph. D. Dissertation, Univers. California, Los Angeles, Chapter 3, 1977
350. S. W. Kirtley: Ph. D. Thesis, Univers. California, Los Angeles, 1972
351. G. Ciani et al.: J. Organomet. Chem. *186*, 353 (1980)
352. G. Ciani et al.: ibid. *170*, C15 (1979)
353. H. B. Chin, R. Bau: Inorg. Chem. *17*, 2314 (1978)
354. L. F. Dahl, J. F. Blount: ibid. *4*, 1373 (1965)
355. F. Takusagawa, A. Fumagalli, T. F. Koetzle: unpublished
356. D. F. Shriver, D. Lehman, D. Strope: J. Am. Chem. Soc. *97*, 1594 (1975)
357. T. V. Ashworth et al.: J.C.S. Chem. Comm. 757 (1977); J.C.S. Dalton 1043 (1978)
358. M. I. Bruce et al.: J.C.S. Chem. Comm. 1041 (1972)
359. A. J. P. Domingos et al.: ibid. 912 (1973)
360. A. W. Parkins et al.: Angew. Chem. Internat. Edit. *9*, 663 (1970)
361. S. Jeannin, Y. Jeannin, G. Lavigne: Inorg. Chem. *17*, 2103 (1978)
362. M. R. Churchill et al.: J. Am. Chem. Soc. *97*, 7158 (1975)
363. B. F. G. Johnson et al.: J. Organomet. Chem. *173*, 187 (1979)
364. a) M. Catti, G. Gervasio, S. A. Mason: J.C.S. Dalton 2260 (1977)
     b) G. Gervasio, G. Ferraris: Cryst. Struct. Comm. *3*, 447 (1973)
365. G. Gervasio, D. Osella, M. Valle: Inorg. Chem. *15*, 1221 (1976)
366. Mentioned in J. Lewis, B. F. G. Johnson: Pure Appl. Chem. *44*, 43 (1975)
367. C. W. Bradford et al.: J.C.S. Chem. Comm. 87 (1972)
368. A. J. Deeming, M. Underhill: ibid. 277 (1973)
369. a) M. R. Churchill et al.: J. Am. Chem. Soc. *98*, 2357 (1976)
     b) M. R. Churchill, B. G. DeBoer: Inorg. Chem. *16*, 1141 (1977)
370. a) J. R. Shapley et al.: J. Am. Chem. Soc. *99*, 8064 (1977)
     b) M. R. Churchill, F. J. Hollander: Inorg. Chem. *17*, 3546 (1978)
371. a) M. R. Churchill et al.: J.C.S. Chem. Comm. 699 (1977)
     b) M. R. Churchill, R. A. Lashewycz: Inorg. Chem. *17*, 1291 (1978)
372. a) J. R. Shapley et al.: J. Organomet. Chem. *162*, C39 (1978)
     b) M. R. Churchill, R. A. Lashewycz: Inorg. Chem. *18*, 848 (1979)
373. B. F. G. Johnson et al.: J.C.S. Chem. Comm. 551 (1978)
374. M. Laing et al.: ibid. 1035 (1978)
375. K. A. Azam et al.: ibid. 1086 (1978)
376. a) L. J. Farrugia et al.: ibid. 260 (1978)
     b) L. J. Farrugia et al.: J.C.S. Dalton, submitted for publication
377. M. R. Churchill et al.: J.C.S. Chem. Comm. 534 (1978)
378. B. F. G. Johnson et al.: J.C.S. Dalton 673 (1978)
379. B. F. G. Johnson et al.: J. Organomet. Chem. *162*, 179 (1978)
380. a) C. R. Eady et al.: J.C.S. Chem. Comm. 602 (1976)
     b) J. J. Guy, G. M. Sheldrick: Acta. Cryst. *B34*, 1718 (1978)
381. S. Bhaduri et al.: J.C.S. Chem. Comm. 343 (1978)
382. R. G. Teller: Ph. D. Dissert. Univers. Southern California, Chapter 3, 1978
383. R. D. Adams, N. M. Golembski: J. Am. Chem. Soc. *101*, 1306 (1979)
384. R. D. Adams, N. M. Golembski: Inorg. Chem. *17*, 1969 (1978)
385. R. D. Adams, N. M. Golembski: ibid. *18*, 2255 (1979)
386. R. D. Adams, N. M. Golembski: J. Am. Chem. Soc. *101*, 2579 (1979)
387. S. Aime et al.: Inorg. Chim. Acta. *32*, 163 (1979)
388. M. R. Churchill, R. A. Lashewycz: Inorg. Chem. *18*, 156 (1979)
389. B. F. G. Johnson et al.: J.C.S. Chem. Comm. 719 (1979)
390. B. F. G. Johnson et al.: J.C.S. Dalton 1356 (1979)
391. A. Nutton et al.: J.C.S. Chem. Comm. 631 (1980)
392. G. Ciani, A. Sironi and S. Martinengo: J. Organomet. Chem. in press
393. L. J. Farrugia et al.: J.C.S. Dalton, submitted for publication
394. L. J. Farrugia et al.: ibid. submitted for publication
395. M. B. Fischer et al.: J. Am. Chem. Soc. *102*, 4941 (1980)
396. S. S. Wreford et al.: ibid. *102*, 1558 (1980)

Spin Crossover in Iron(II)-Complexes

Philipp Gütlich

Institut für Anorganische und Analytische Chemie, Universität, D-6500 Mainz

List of Symbols

HS	high-spin	T_c	transition temperature
LS	low-spin	μ	magnetic moment
$\bar{P}$	mean spin-pairing energy	δ	Mössbauer isomer shift
Δ	cubic ligand field parameter	ΔE_Q	quadrupole splitting
k_B	Boltzmann factor		

Ligand Abbreviations

phen	1,10-phenanthroline	medpma	di-(2-pyridylmethyl)methyl-amine
bipy	2,2'-bipyridine		
ox	oxalate	ppa	N'-(2-pyridylmethyl)picolyl-amidine
mal	malonate		
py	pyridine	pnp	2,6-bis-(2-diphenylphosphino-ethyl)pyridine
2-pic	2-picolylamine		
pi	α-picoline	pmp	2,6-bis-(2-diphenylphosphino-methyl)pyridine
pip	2-pyridinaldehyde-N-iso-propylimine		
		HB(pz)$_3$	hydrotris(1-pyrazolyl)borate
pmi	2-pyridinaldehyde-N-methyli-mine	btz	2,2'-bi-4,5-dihydrothiazine
		bt	2,2'-bi-2-thiazoline
pyim	2-(2'-pyridyl)imidazole	dippen	cis-1,2-bis(diphenylphosphino)-ethylene
pyimi	2-(2'-pyridyl)imidazoline		
pyben	2-(2'-pyridyl)benzimidazole	tren	tris(2-aminoethyl)amine
ppp	3-phenyl-5-pyridyl-(2)-pyrazole	[14]dieneN$_4$	5,7,7,12,14,14-hexamethyl-1,4,8,11-tetraazacyclotetra-deca-4,11-diene
ppi	2-pyridinalphenylimine		
pythiaz	2,4-bis(2-pyridyl)thiazole		
paptH	2-(2-pyridylamino)-4-(2-pyridyl)thiazole	[14]aneN$_4$	mesa-5,5,7,12,12,14-hexa-methyl-1,4,8,11-tetraazacyclo-tetradecane
dbtp	2,6-(dibenzothiazol-2-yl)-pyridine	dtc	dithiocarbamate
terpy	2,2',2''-terpyridyl	tfd	trifluoromethyldithiolene
dpma	di-(2-pyridylmethyl)amine	mnt	maleonitriledithiolene

1 Introduction

Transition metal complexes with d^4, d^5, d^6, and d^7 electron configurations usually are capable of forming either spin-free (high-spin, HS) or spin-paired (low-spin, LS) ground states, depending on the strength of the cubic ligand field, Δ, relative to the mean spin-pairing energy, $\bar{P}$. The simple inequality [1].

$$\Delta(HS) < \bar{P} < \Delta(LS) \tag{1.1}$$

expresses the conditions for finding a particular transition metal complex with either HS behaviour (complexes with weak ligand fields $\Delta < \bar{P}$) or LS behaviour (complexes with strong ligand fields $\Delta > \bar{P}$). Intermediate-spin (IS) ground states are rarely en-

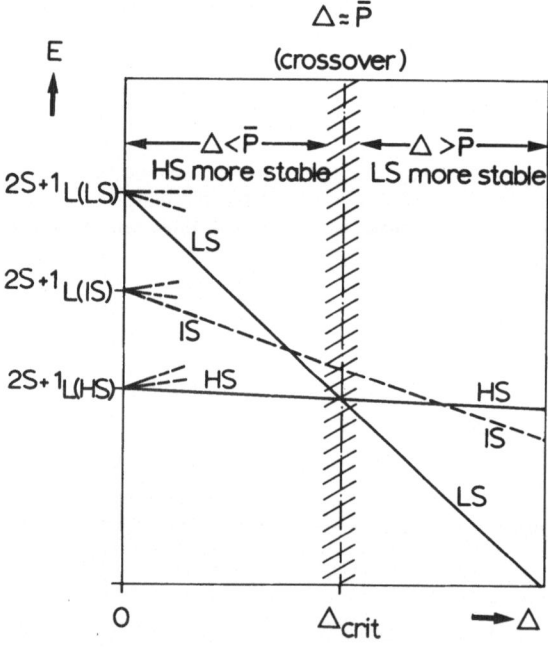

Fig. 1. Schematic energy level diagram for crystal field states as function of the ligand field strength Δ. At $\Delta = 0$ (free ion case), the coupling between the valence electrons of the transition metal ion yield the pure Russell-Saunders terms ^{2S+1}L, with the high-spin (HS) state being lowest by virtue of Hund's first rule. At $\Delta > 0$ (ligands attached to metal ion) the Russell-Saunders terms split into crystal field states $^{2S+1}$, the number and irreducible representation of which are determined by the molecular symmetry [2,3]. For $\Delta < \bar{P}$ ($\bar{P}$, mean spin pairing energy), the HS crystal field state is ground state. At a critical field strength Δ_{crit}, $\bar{P}$ becomes equal to Δ and the HS and the LS (low-spin) crystal field states cross. At $\Delta > \bar{P}$, the LS crystal field state is more stable (contrary to Hund's first rule). Crystal field ground states of intermediate-spin (IS) are rarely encountered. In the region where Δ becomes comparable to $\bar{P}$ ($\Delta \approx \bar{P}$, shadowed area), thermally induced high-spin low-spin transition is possible, provided the energy difference $|\Delta - \bar{P}|$ is on the order of kT

countered; only a few examples have been reported, e.g. iron(II) complexes with strong axial distortions. The situation is sketchily illustrated in Fig. 1.

By far the largest portion of transition metal compounds described in the literature have either HS or LS ground states, whose magnetism follows the Curie or Curie-Weiss law. Relatively few transition metal compounds, all of them containing first-row transition metal ions, have ligand field strengths in the close proximity of the mean spin pairing energy. In such systems $|\Delta - \bar{P}|$ may become comparable to $k_B T$ with the result that both spin states HS and LS may be populated and coexist in thermal equilibrium. The fractional ratio of the different spin states depends on temperature and pressure under otherwise constant conditions. This phenomenon is commonly referred to as *spin crossover, magnetic crossover, spin transition*, or *spin equilibrium.*

The first examples exhibiting a temperature dependent conversion between different spin states have been found by Cambi and his school[4] in the early thirties while studying trisdithiocarbamates of iron(III) with various N-substituted ligands. By now many more spin crossover systems have been discovered, particularly in the complex chemistry of iron(II), iron(III), and cobalt(II); a few examples of nickel(II) and manganese(II) compounds have been added recently.

Various types of the spin crossover characteristics, expressed in terms of the fraction x_{HS} of molecules in the HS state as a function of temperature, have been observed (see Fig. 2):

(i) "Abrupt" transition within a few Kelvin as, for instance, in $[Fe(phen)_2(NCS)_2]$[5], or "gradual" transition extended over a large temperature range as, for instance, in $[Fe(2-CH_3-phen)_3]X_2$ [89,102] (see Fig. 2a);

(ii) transitions which are incomplete on either the low temperature end, merging into a plateau of what is called "residual paramagnetism" (RP), or in the high temperature region (see Fig. 2b);

(iii) spin transitions showing hysteresis of the width $T = T_c^> - T_c^<$ (see Fig. 2c).

The phenomenon of magnetic crossover has been observed not only in the solid state, but to a lesser extent, *it has also been followed in solution* as exemplified by the recent reports from Wilson's group. In fact, solution state measurements on crossover systems have turned out to be most appropriate in studying the effect of ligand substitution on the magnetic behaviour; unpredictable influences originating from lattice forces are excluded this way.

The easiest way of following the spin conversion as a function of temperature is the measurement of the magnetic susceptibility. Mössbauer spectroscopy, on the other hand, has proven to be an extremely powerful tool in following spin crossover in iron complexes. The reason is obvious: whereas the $\chi(T)$ curve reflects the overall change in magnetism of the bulk material, the Mössbauer spectrum presents resonance lines of the individual spin states separately, provided the lifetime of the particular spin states are comparable with or longer than the Mössbauer spectroscopy time scale, which is in case of ^{57}Fe spectroscopy determined by the quadrupole precession time of the $I = 3/2$ state ($\sim 10^{-7}$ s). There are other techniques which have been extremely valuable in detecting the changes of certain physical properties, associated with a temperature (or pressure) induced spin transition, such as: the lattice parameters by X-ray diffraction, electronic d-d transition, and metal $\rightleftharpoons$ ligand charge

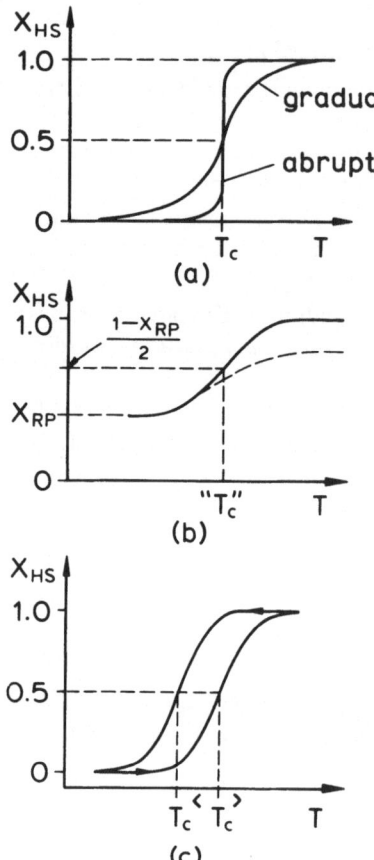

Fig. 2a–c. Schematic representation of various types of spin crossover behaviour. x_{HS} (T) is the fraction of high spin molecules as a function of temperature. **a** Gradual and abrupt spin transition, respectively; **b** incomplete spin transition (RP is the residual fraction of high spin molecules); **c** spin transition with hysteresis, $T_c^<$ and $T_c^>$ are the transition temperatures in the cooling and heating mode, respectively

transfer by UV-vis spectroscopy, the change of the g-tensor by EPR spectroscopy, intramolecular vibrations by IR and FIR spectroscopy, thermodynamic quantities by heat capacity measurements, specific hyperfine interactions by NMR and ESR spectroscopy, the redox behaviour in solution by electrochemical studies, the lifetime of the nearly equienergetic spin states in solution by ultrasonic and temperature-jump relaxation measurements; in very few instances have ESCA and X-ray absorption edge measurements successfully been employed in spin crossover research. All these methods will be referred to in more detail later in this review.

It is by now well documented that the spin transition behaviour, i.e. the molar fractions of the different spin states at a given temperature, the amount of residual paramagnetism at low temperature, the transition temperature T_c (= temperature of equal amounts of the coexisting spin states), the shape of the conversion curve (more or less gradual), eventual hysteresis, may be influenced by various factors, most important among these being

— intraligand substitution
— ligand replacement

- nature of non-coordinating solvent
- nature of crystal solvent
- isotypic metal dilution
- hydrogen bonding
- sample preparation
- solute-solvent interaction.

Beside the accumulation of a large amount of experimental data considerable efforts have been undertaken in trying to put experimental facts on theoretical grounds. The goal here can be considered twofold:

(i) Finding optimal congruency between the experimentally determined spin conversion curve, e.g. $\chi(T)$, and a model based on a Boltzmann distribution of the electrons over all thermally accessible states under the inclusion of various kinds of supposedly significant perturbations (e.g. spin-orbit coupling, crystal field distortion, configuration interaction, covalency effects);

(ii) trying to find a model, based on ideas which could eventually be crucial in the mechanism of spin transition processes, which is capable of explaining abrupt as well as smooth spin transitions, the residual paramagnetism at low temperatures, and hysteresis effects. Various theoretical accounts on spin transition phenomena reported in the literature will be referred to at the end of this review.

Research activities in the field of spin crossover in transition metal chemistry have been reviewed by a number of authors [6-15].

Lindoy and Livingstone [6] present a discussion on complexes of iron(II), cobalt(II), and nickel(II) with α-diimines and related bidendate ligands, some of which have been found to induce temperature dependent spin transitions. Barefield, Busch, and Nelson [7] review representative examples of those classes of compounds of iron(II), cobalt(II), and nickel(II) which exhibit anomalous magnetic moments. König [8] has focussed his interest on octahedral bis(2,2'-bipyridyl) and bis(1,10-phenanthroline) complexes of iron(II) and covers well the results of physical measurements on $[Fe(phen)_2 X_2]$ and related compounds with pure S = 2, S = 1, and S = 0 ground states and those with $^5T_2(O_h) \rightleftharpoons {}^1A_1(O_h)$ crossover. An excellent review has been given by Martin and White [9], particularly suited for those readers with no background knowledge on spin crossover. The authors start out with a brief ligand-field theoretical description of the crossover situation and continue with a discussion of the factors affecting the electronic crossover such as the effect of chemical modification of the ligands, the effect of temperature and pressure, solid state effects, crystal field distortions and spin-orbit coupling, configuration interaction, finally followed by a survey upon the physical properties of some spin-crossover systems as reflected in the magnetic behaviour, Mössbauer effect, NMR, UV-vis, and IR spectra, and phase changes. Sacconi's lecture on "Conformational and Spin State Interconversions in Transition Metal Complexes" concerns five-coordinate and six-coordinate complexes of iron(II), cobalt(II) and nickel(II), which have been found to exhibit various spin multiplicities in thermal equilibrium [10]. König [11], in a plenary lecture on spin crossover, interprets energy level diagrams on the basis of ligand field theory, with main emphasis on the crossover region for $^5T_2(O_h) \rightleftharpoons {}^1A_1(O_h)$ transitions, and presents a brief review of the most important experimental results on spin transitions in iron(II) complexes, and finally mentions various possible mechanisms. In a somewhat later account [13], König and Ritter demonstrate the extraordinary usefulness of ^{57}Fe Mössbauer spectroscopy in spin crossover studies of iron compounds. The most complete review on spin transitions in six-coordinate iron(II) complexes has been presented by Goodwin [14]. Aspects of spin transition have also been reviewed by Machado [16]. Drickamer and Frank [17] have summarized the work on pressure-induced spin transitions in iron complexes. Results and mechanistic ideas from recent Mössbauer effect measurements on spin crossover systems of iron(II) have been communicated by the present author [15].

The present review deals with spin crossover in iron(II) complexes only. It was intended to cover the relevant literature rigorously up to early 1980. The author wishes to apologize in case one or the other important publication slipped his attention. It is planned that a similar account on spin crossover systems of other 3 d ions follows later in this series.

2 Conditions for Spin Crossover

It is well established that ligand field theory provides a suitable basis for the interpretation of the magnetic properties of transition metal complexes [2,3,18-21]. Complexes containing σ- and π-donor ligands, so-called weak ligands in the sense of the spectrochemical series such as the halide ions, water, hydroxyl, oxalate, have cubic ligand field strengths $\Delta(HS)$ well below the mean spin-pairing energy $\overline{P}$. Thus Hund's first rule of maximum spin multiplicity is obeyed at all temperatures and the spin state remains the same. So-called strong ligands with σ-donor and π-acceptor capability such as CN^-, CO, NO^+, phosphines, arsines, may exert field strengths $\Delta(LS)$ well above $\overline{P}$. The spin state will be minimal in such complexes and independent of the temperature. A theoretical description of the temperature dependence of the magnetic susceptibility of both classes of compounds, employing van Vleck's theory [22], generally imposes no principal difficulties. Successful calculations should include perturbations of the energy levels by axial crystal field components and spin-orbit coupling as well as configuration interaction and covalency effects.

The spin state of a complex molecule changes from one spin multiplicity to another, say from $^5T_{2g}$ to $^1A_{1g}$ for a d^6 cation in the approximation of O_h symmetry, if the ligand field strength exceeds the critical value of $\Delta_c = \overline{P}$. At this "crossover point" the change in ligand field energy associated with the electron transfer between the configurations of different spin multiplicity is exactly outweighted by the sum of the changes in coulombic repulsion (P_c) and quantum mechanical exchange energy (P_e). Approximate values of P_c, P_e, and $\overline{P}$ for first-row transition metal ions Me^{2+} and Me^{3+} with octahedral symmetry have first been estimated and expressed in terms of Racah's parameters B and C by Orgel [23] and Griffith [24] independently from each other. The spin-pairing energy has further been dealt with by Jørgensen [25], who introduced a spin-pairing energy parameter D in terms of Racah's parameters of interelectronic repulsion. König and Schläfer [26] have derived a value of $\overline{P} \approx$ 11 900 cm^{-1} from electronic spectra of $[Fe(phen)_3]^{2+}$ compounds. This value is about one third smaller than the free ion value for $Me^{2+}(d^6)$ ions from Griffith and

Orgel, which implies that the Racah parameters are considerably reduced by complex formation. The metal d-electrons delocalize towards the ligands and thereby lower the interelectronic repulsion[27,28]. Exact spin-pairing energies (π) at the crossover points in octahedral d^4, d^5, d^6, and d^7 transition metal complexes have been calculated by König and Kremer[29] by direct diagonalization of the relevant ligand field plus interelectronic repulsion matrices. Their results are presented in terms of π/B as function of C/B. π differs significantly from the earlier $\overline{P}$ data, which have been found without taking electron configuration into account. The authors provide also estimates of spin-pairing energies on the basis of magnetic and electronic spectral data.

A direct determination of the critical field strength at the crossover point is not possible. The reason is that $\Delta(HS)$ and $\Delta(LS)$ in the general condition for the existence of (nearly) equienergetic HS and LS groundstates[1], $\Delta(HS) < \overline{P}(\pi) < \Delta(LS)$, never become exactly equal. As soon as the individual complex molecule changes spin from LS to HS or vice versa, the intramolecular lattice parameters change accordingly, which has been found in various temperature dependent X-ray diffraction studies[30,31] and from the pressure dependence of the magnetism of spin crossover systems in solution[1,32]. It is obvious that the spin state induced changes of intramolecular volume and along with it the vibrational characteristics (see for instance [33–35] and other work referred to later in this article) simultaneously alter the ligand field strength. The true critical field strength $\Delta_c = \overline{P}(\pi)$ may be considered that of an activated complex to be passed through during the relaxation between the different spin states. One way of approaching Δ_c could be trying to extract $\Delta(HS)$ and $\Delta(LS)$ values from temperature dependent UV-vis spectra. This, however, fails in general, because the weak d-d transitions are in most cases, particularly in low-spin iron(II) compounds, obscured by very strong charge-transfer bands. MCD spectroscopy eventually could help to improve the situation. Some authors have persued a different way to arrive at reasonable estimates of Δ_c: Robinson et al.[36] have studied the electronic spectra of octahedral nickel(II) complexes (high-spin) with ligands of the α-diimine and closely related classes and found, by comparison with corresponding iron(II) complexes of both high-spin and low-spin types, that the spin crossover in iron(II) complexes should occur at $\Delta_c = \overline{P} = 12\,500 \pm 800 \text{ cm}^{-1}$. Wilson et al.[37] have followed the same method in an electronic spectral study of some iron(II) crossover complexes with hexadentate ligands in solution. By comparing spectral and structural data for the iron(II) complexes with those of their nickel(II) analogs it has been possible to establish the critical $\Delta_{HS}(Fe^{2+})/\Delta(Ni^{2+})$ ratio to be 1.07 and that of $\Delta_{LS}(Fe^{2+})/\Delta(Ni^{2+})$ to be 1.15. Further, in good agreement with Robinson et al.[36], they evaluated the mean spin-pairing energy $\overline{P}$ to be $12\,800 \pm 400 \text{ cm}^{-1}$ for the iron(II) complexes with hexadentate ligands under study.

The spectral parameters $\Delta(^5T_2) = 11\,900 \text{ cm}^{-1}$, $B(^5T_2) \approx 640 \text{ cm}^{-1}$ for the high-spin state and $\Delta(^1A_1) \approx 16\,300 \text{ cm}^{-1}$, $B(^1A_1) \approx 580 \text{ cm}^{-1}$ for the low-spin state were estimated by König and Madeja[38] from the d-d transitions in the spin crossover systems $[Fe(phen)_2(NCS)_2]$ and $[Fe(phen)_2(NCSe)_2]$. They have ascribed the relatively large difference between $\Delta(^5T_2)$ and $\Delta(^1A_1)$ to significant changes accompanying the change in the spin state.

For comparison, in iron(III) crossover systems the critical field strength is conceivably higher in energy than in iron(II) complexes. As an example, Tsipis et al.[39]

have estimated Δ_c to be about $16\,500 \text{ cm}^{-1}$ for tris(di-n-propyldithiocarbamato) iron(III).

The relative stability of the different spin states constituting the equilibrium

$$\text{LS} \rightleftharpoons \text{HS} \tag{2.1}$$

is determined by the free energy change

$$\Delta G = G(\text{HS}) - G(\text{LS}) = \Delta H - T\,\Delta S. \tag{2.2}$$

Naturally, at absolute zero the entropy term $T\,\Delta S = T[S(\text{HS}) - S(\text{LS})]$ vanishes, and the free energy change ΔG is identical to the enthalpy change $\Delta H = H(\text{HS}) - H(\text{LS})$. At sufficiently low temperatures the enthalpy term in Eq. (2.2) still dominates and a discussion on the relative stability of coexisting ground states may ignore the entropy term. ΔH and ΔS may be composed of various contributions:

$$\Delta H = \Delta H_{el} + \Delta H_{intravib} + \Delta H_{lat.vib} \tag{2.3}$$

$$\Delta S = \Delta S_{el} + \Delta S_{intravib} + \Delta S_{lat.vib}. \tag{2.4}$$

The subscript notations "el" refers to the electronic enthalpy (entropy) change associated with the spin change; "intravib" and "lat.vib" denote contributions from changes in intramolecular and lattice vibrational characteristics accompanying the spin transition. In case of a LS $\rightleftharpoons$ HS transition ΔH_{el} should be approximately equal to the energy separation between the lowest spin-orbit crystal field state of the LS and that of the HS multiplet, respectively. Some minor correction may be caused by electron-phonon interaction. It is well established that $\Delta H_{el} > 0$ on going from the LS to the HS state. ΔH_{el} generally being the largest contribution in Eq. (2.3) ($\Delta H_{el} \approx 10^3 \text{ cm}^{-1}$, $\Delta H_{intravib} \approx 10^2 \text{ cm}^{-1}$, and $\Delta H_{lat.vib} \approx 10 \text{ cm}^{-1}$ to give some orders of magnitude for the differences) implies that $\Delta H = H(\text{HS}) - H(\text{LS}) > 0$. We also find that $\Delta S > 0$ on going from the LS to the HS state. ΔS_{el} comprises contributions from the differences in the spin and orbital degeneracies,

$$\Delta S_{el}^{spin} = R[\ln{(2\,S + 1)_{HS}} - \ln{(2\,S + 1)_{LS}}] \tag{2.5}$$

$$\Delta S_{el}^{orb} = R[\ln{(2\,L' + 1)_{HS}} - \ln{(2\,L' + 1)_{LS}}] \tag{2.6}$$

where $(2\,L' + 1)$ adopts the values of 3, 2, and 1 for T, E, and A or B states, serially. ΔS_{el}^{spin} is always positive. ΔS_{el}^{orb} may in principle be positive and negative (as in case of a $^6A_1(O_h) \rightleftharpoons {}^2T_2(O_h)$ spin transition). In actual cases, however, spin crossover complexes have lower than cubic symmetry so that orbital degeneracy is partially or (more often) completely removed and ΔS_{el}^{orb} is no longer important. Most important, however, is the entropy contribution from changes in the vibrational behaviour of the crossover system, predominantly the metal-ligand stretching and deformation vibrations with only a minor contribution from the lattice modes. This has been pointed out by Sorai and Seki, who found by precise heat capacity measurements on the spin transition systems [Fe(phen)$_2$(NCS)$_2$] and [Fe(phen)$_2$(NCSe)$_2$], that only about one third of the total entropy change ΔS of nearly 50 J K^{-1}, on passing from the LS to the HS state, stems from ΔS_{el} and about two thirds originate from the changes in phonon modes with about 50 per cent of the total ΔS_{vib} being accounted for by changes in metal-ligand stretching frequencies[34].

It is apparent from Eq. (2.2) that, as long as the temperature is sufficiently small such as to yield $\Delta H > T \Delta S$, the LS state is more stable than the HS state and the equilibrium, Eq. (2.1), is shifted towards the LS side. At the transition temperature T_c, ΔH and $T \Delta S$ are equal in magnitude and thus $\Delta G = 0$. Both spin isomers are equienergetic (in free energy and not in enthalpy terms!) and should be present in equal amounts. Above the transition temperature the HS state becomes more and more favoured with increasing temperature due to the dominating entropy term causing $\Delta G < 0$. Thus the driving force for the spin transition from LS to HS is the gain in entropy.

3 Bis(phenanthroline) and Bis(bipyridine) Complexes

3.1 FeL$_2$X$_2$ Complexes with Unsubstituted Phenanthroline and Bipyridine Ligands (L = phen, bipy)

It has been known for a long time that the family of the tris(phenanthroline) iron (II) complexes [Fe(phen)$_3$]X$_2 \cdot$ nH$_2$O is low-spin with an $^1A_1(O_h)$ ground state, independent of the nature of the counterion X and the amount of crystal water[40–45]. The replacement of one phen ligand by either two monodentate ligands X or a different bidentate ligand X$_2$ changes the magnetism of the iron (II) complex drastically, and one can distinguish between four classes:

a) Low-spin complexes with a spin singlet ground state, $^1A_1(O_h)$[1]
$\quad$ X = CN$^-$ (2 H$_2$O)[2] [46–50]
$\quad$ X = CNO$^-$ [51]
$\quad$ X = NO$_2^-$ [52]
b) High-spin complexes with a spin quintet ground state, $^5T_2(O_h)$
$\quad$ X = Cl$^-$ [5,45,47,48,49,50,53,54,55,56,57]
$\quad$ X = Br$^-$ [5,47,48,49,50,54,55,56,57]
$\quad$ X = I$^-$ [5,47,49]
$\quad$ X = N$_3^-$ [5,48,49,50,54,56,57]
$\quad$ X = NCO$^-$ [45,48,50,54,56,57]
$\quad$ X = HCOO$^-$ [45,48,54,56,57]
$\quad$ X = CH$_3$COO$^-$ [50,54,56]
$\quad$ X = $\frac{1}{2}$ mal (2 H$_2$O)

1 $\quad$ For convenience, the ground state notations for the approximation of O$_h$ symmetry are used; the actual symmetry, of course, is lower than O$_h$
2 $\quad$ The amount of crystal water is given in parentheses

c) Intermediate-spin complexes with a spin triplet ground state, $^3T_1(O_h)$

$X = F^- \ (4\,H_2O)^{56,59)}$

$X = \frac{1}{2}\,ox\ (5\,H_2O)^{56,59-61)}$

$X = \frac{1}{2}\,mal\ (7\,H_2O)^{56,59,60)}.$

A recent Mössbauer and magnetic susceptibility study on [Fe(phen)$_2$ox]·5 H$_2$O[186)] has identified this complex as a typical $^5T_2(O_h) \rightleftharpoons {}^1A_1(O_h)$ spin crossover system rather than a complex with a spin triplet ground state. Moreover, it has been shown[186)] that the magnetic properties are very susceptible to the amount of lattice water present: Pumping off the lattice water gradually favours more and more the HS state; the stable monohydrate, finally, turned out to be a paramagnet.

d) Spin crossover systems, $^5T_2(O_h) \rightleftharpoons {}^1A_1(O_h)$

$X = NCS^{-\ 5,33,34,50,56,58,62-70)}$

$X = NCSe^{-\ 5,34,56,58,64,66,68)}$

In all instances the spin state of the ground state has been established by measuring the magnetic susceptibility and deriving the magnetic moment. Magnetic data of the complexes listed above are given in[8)]. Various other methods have also been used here to characterize the ground state and molecular properties: IR(FIR) spectroscopy[33,34,45,50,58,60,62,64,66,68)], UV-vis spectroscopy[45,48,62,64)], Mössbauer spectroscopy[45,49,55,56,60-64,69)], X-ray investigation[5,62)] heat capacity[34)], proton magnetic resonance spectroscopy[63)]. Clearly, the change in the spin state is a direct consequence of the alteration of the ligand field strength as a function of the σ- and π-bonding properties of the ligand X. On following the complexes [Fe(phen)$_2$X$_2$] in the order of high-spin (S = 2), intermediate-spin (S = 1), spin crossover (S = 2 $\rightleftharpoons$ S = 0), and low-spin (S = 0) one finds the same ordering for the ligands as in the spectrochemical series.

2,2'-bipyridine (bipy) has been found to have very similar bonding properties as 1,10-phenanthroline. The [Fe(bipy)$_3$]$^{2+}$ complex ion is low-spin with ~ 1 B.M. and a $^1A_1(O_h)$ ground state[41,42)]. The magnetism of the complex [Fe(bipy)$_2$X$_2$]·nH$_2$O parallels that of the [Fe(phen)$_2$X$_2$]·nH$_2$O compounds. The same four classes of compounds as above are encountered for [Fe(bipy)$_2$X$_2$]·nH$_2$O, depending on the partial ligand field strength exerted by the ligand X. [Fe(bipy)$_2$(CN)$_2$]·3 H$_2$O with the strong π-backbonding CN$^-$ ligand is low-spin as the parent system[46,49)]. The paramagnetic [Fe(bipy)$_2$Cl$_2$] is a member of the class b) compounds with a $^5T_2(O_h)$ ground state[56)]. [Fe(bipy)$_2$Ox]·3 H$_2$O and [Fe(bipy)$_2$mal]·3 H$_2$O have spin-triplet ground states, $^3T_1(O_h)$[56,59)]. Finally, the system [Fe(bipy)$_2$(NCS)$_2$] shows a temperature dependent $^5T_2(O_h) \rightleftharpoons {}^1A_1(O_h)$ transition[5,71,72)], with a sharp drop in the magnetic moment around 215 K.

The following three crossover systems

I: [Fe(phen)$_2$(NCS)$_2$]

II: [Fe(phen)$_2$(NCSe)$_2$]

III: [Fe(bipy)$_2$(NCS)$_2$]

are among those which have been studied most extensively using various techniques.

3.1.1 Magnetism

Baker and Bobonich[5] first observed an anomalous temperature dependence of the magnetic moment for I, II, and III, which decreases sharply from 5.0 B.M. to 1.5 B.M. around 175 K in case of I, at $\sim$ 235 K in case of II, and at $\sim$ 215 K in case of III. In a more refined study of the susceptibility between 440 K and 77 K, König and Madeja[62,64] have confirmed the anomalous magnetic behaviour of I and II. They measured a limiting high-temperature magnetic moment of $\sim$ 5.20 B.M. at 440 K for both systems. They further observed a sharp drop of the magnetic moment (within a few degrees Kelvin) at T_c $\sim$ 174 K for I and at T_c $\sim$ 235 K for II; the low-temperature moments were found to approach 0.65 B.M. in case of I, and $\sim$ 0.84 B.M. in case of II. They attempted to fit a van Vleck type expression to the $\chi(T)$ data, considering a Boltzmann distribution over the $^1A_1$ ground state and the fifteen $^5T_2(O_h)$ spin-orbit levels, whose center of gravity was estimated to be separated from the $^1A_1(O_h)$ state by $\sim$ 350 cm^{-1}. This procedure failed to predict the sharp change of the magnetic moment. Most interesting is their observation that the magnetism, particularly the fraction of the "residual paramagnetism" at low temperatures, depends markedly on the method of preparation (see Fig. 3), although elemental

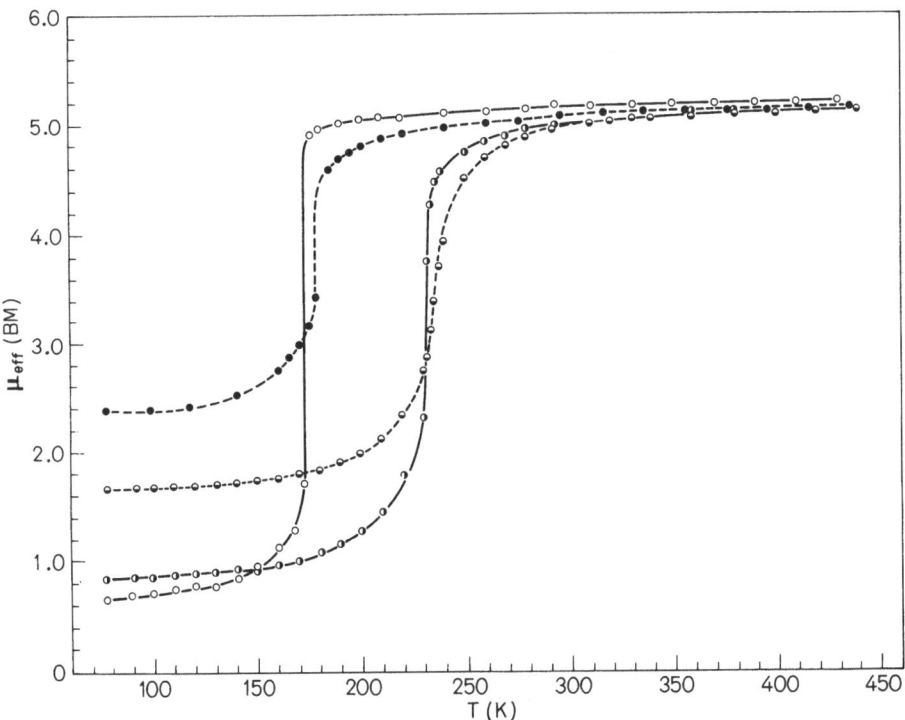

Fig. 3. Magnetic moment vs. temperature for [Fe(phen)$_2$(NCS)$_2$] ($\circ$, preparation A; $\bullet$, preparation B) and for [Fe(phen)$_2$(NCSe)$_2$] ($\circ$, preparation A; $\bullet$, preparation B) showing the influence of the method of sample preparation on the magnetism (from Ref. 64)

analysis, Mössbauer effect measurements, IR and electronic spectra, and X-ray powder photographs did not reveal any apparent differences for the samples obtained by different methods (method A: extraction of one phen ligand from I with ace-tone [54]; method B: adding a solution of phen in pyridine to a solution of $[Fe(py)_4(NCS)_2]$ in dry pyridine).

An interesting observation has been reported on by Casey and Isaac [65], who noticed a certain time dependence of the magnetic susceptibility in polycrystals of I. Setting the cryostat near the transition temperature ($T_c \sim 174$ K), a period of up to two hours was required to complete the transition. This implies an appreciable energy barrier, and the authors conclude that there might be a significant rearrange-ment of the environment of the iron atom, and that therefore, contrary to earlier statements drawn from X-ray powder patterns [5,64], a crystallographic phase change cannot be ruled out without a detailed structural analysis.

The magnetism of III, $[Fe(bipy)_2(NCS)_2]$, has also been found to vary distinctly with the method of preparation [71]. Three polymorphs of III have been obtained by slightly modified procedures. They all differed in their $\mu(T)$ curves, particularly in the low-temperature region, where the three polymorphs showed limiting magnetic moments of 0.94 B.M., 1.39 B.M., and 1.61 B.M., respectively, at 77 K. No signifi-cant change, however, was observed in the elemental analysis, in the IR and the Möss-bauer spectra of the three polymorphs. The differences in the low-temperature limits of the magnetic moments indicate that the transition from $^5T_2(O_h)$ to $^1A_1(O_h)$ is more or less complete but a small fraction of molecules remains in the $^5T_2$ state in the low-temperature region, provided the presence of paramagnetic impurity can be ruled out, which has been shown for the polymorph with 0.94 B.M. (77 K) by means of Mössbauer spectroscopy [71].

The existence of three polymorphs of III has been confirmed by Casey [72], who has carried out seventeen preparations of the compound by two different methods and measured the magnetic susceptibility over a temperature range of 300—85 K. Within the experimental error, he found the same values for the transition tempera-ture (212 ± 3 K) and the low-temperature magnetic moments (0.91, 1.35, and 1.65 B.M., respectively) for the three polymorphs as in Ref. 71. He further observed that, as with I [65], the change from the HS state to the LS state at the transition temperature (~ 213 K) took on the order of two hours to complete, and therefore concluded that a simple thermal equilibrium between the two spin states may be ruled out.

Madeja [52] has claimed that there exists a red modification of III with an S = 1 ground state. This was questioned by Driver and Walker [73], but finally confirmed by Casey and Thackeray [74].

Two novel complexes of composition $[Fe(bipy)_7(NCS)_6]$ (A) and $[Fe_3(bipy)_7(NCSe)_6]$ (B) obtained by abstraction of bipy from $[Fe(bipy)_3]X_2$, X = NCS^-, $NCSe^-$ have been reported [75]. The magnetic moments are intermediate between those of $^5T_2(O_h)$ and $^1A_1(O_h)$ ground state, viz. 3.23 and 3.24 B.M. at 294 K for A and B, respectively. The susceptibility follows the Curie-Weiss law between 77 and 294 K. These findings, together with the recorded ^{57}Fe Mössbauer spectra, (which yielded two overlapping spectra with parameter values which are characteristic of high-spin iron (II) and low-spin iron (II), respectively, and with intensities independent of

temperature), led the authors to conclude that both systems A and B contain constant
2:1 ratios of low-spin $^1A_1(O_h)$ and high-spin $^5T_2(O_h)$ species. Thus a high-spin $\rightleftharpoons$
low-spin transition does not occur here.

The system I has been reported to crystallize with different amounts of water,
and this apparently has a surprisingly dramatic influence on the magnetic properties:
The dihydrate $[Fe(phen)_2(NCS)_2] \cdot 2\,H_2O$ shows low-spin behaviour with μ_{eff} ranging
from 2.97 B.M. at 298 K to 2.80 B.M. at 88 K. The monohydrate which was prepared by
two different methods (yielding no major difference in magnetism), still shows a
temperature dependent spin transition with a rather sharp change of μ_{eff} near 180 K[67].

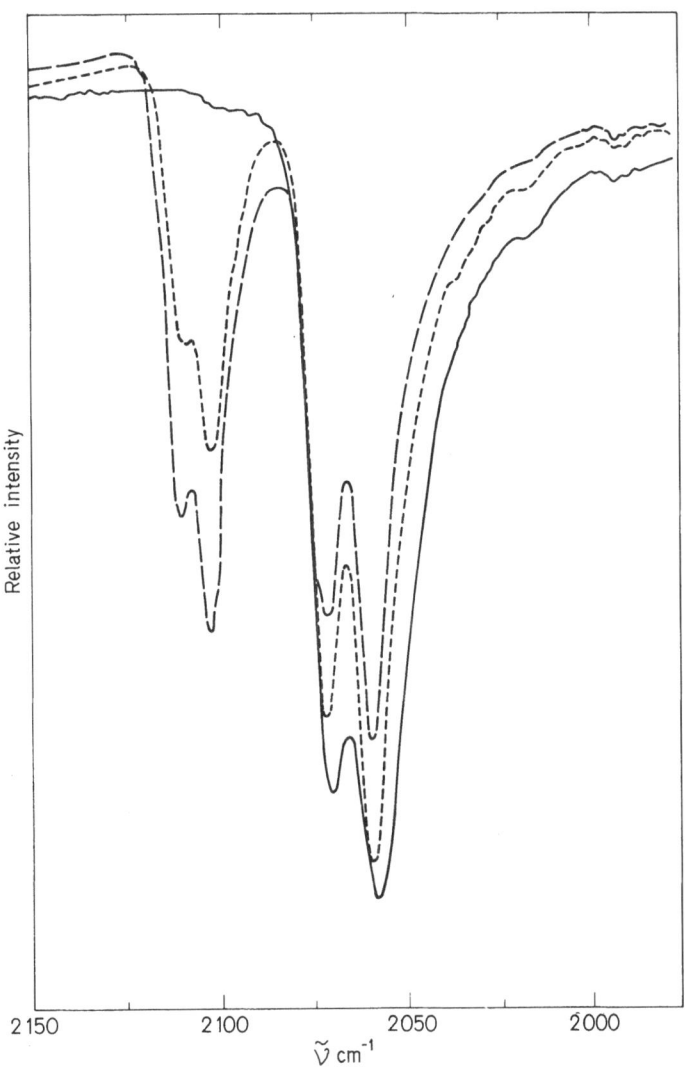

Fig. 4. IR spectra of $[Fe(phen)_2(NCS)_2]$ at room temperature (———), 172 K ($\cdots\cdots$), and
160 K (————) (from Ref. 58)

3.1.2 Vibrational Properties

Baker and Long[58] have measured the infrared spectra of I, II, and III at various
temperatures between 300 and 77 K. They focussed their interest primarily on the
C–N stretching band of NCS$^-$ and NCSe$^-$, which they found at room temperature
as a strong doublet at 2060–2070 cm^{-1}. On cooling, the intensity of this doublet
decreased in favour of a new doublet which appeared at 2100–2110 cm^{-1} (see
Fig. 4). The doublet at 2060 cm^{-1} was assigned by the authors to the high-spin
($^5T_2$) isomer, and that at 2100 cm^{-1} to the low-spin ($^1A_1$) isomer. Fig. 5 shows the
relative intensity of the 2100 cm^{-1} doublet of I, II, and III as a function of temperature.

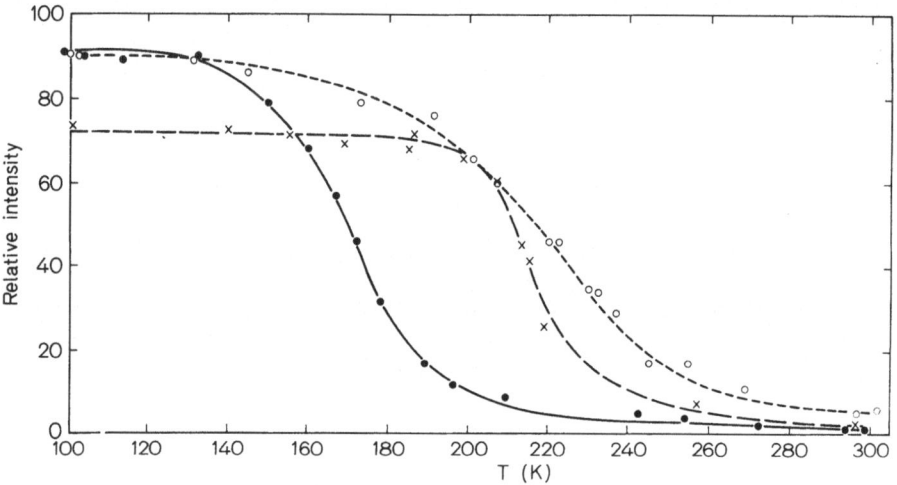

Fig. 5. Relative intensity of the 2100 cm^{-1} doublet vs. temperature for [Fe(bipy)$_2$(NCS)$_2$]
(---), [Fe(phen)$_2$(NCS)$_2$] (——), and [Fe(phen)$_2$(NCSe)$_2$] (-----) (from Ref. 58)

    In another temperature dependent IR study (at room temperature and at
~ 105 K), König and Madeja[64,66] have found essentially the same vibrational
behaviour for I and II as Baker and Long[58], viz. an increase of the C–N stretching
frequency of the NCS$^-$ and NCSe$^-$ ligand by ca. 40 cm^{-1} on going from room
temperature ($^5T_2$ state) to ~ 105 K ($^1A_1$ state). This frequency shift indicates a
strengthening of the C–N bond. The authors explained this as being due to an
extensive increase in π-back donation in the Fe–N(phen) bond, thereby increasing
the Fe–N(phen) bond strength considerably, on going from the high-spin to the low-
spin state, and a sumiltaneous electron delocalization from the C–N bond to make
up for the t$_2$ electron deficiency at the iron ion caused by the enhanced metal →
ligand π-back donation in the Fe–N(phen) bond.
    The same arguments were used to explain the shift of ca. 50 cm^{-1} to higher
frequencies in the C–N stretching mode of III, when passing from the high-spin to
the low-spin state[71].

Table 1. Fe-N(ligand) stretching vibrations ($\bar{\nu}$ in cm^{-1}) in $[Fe(phen)_2(NCS)_2]$, $[Fe(phen)_2(NCSe)_2]$ and $[Fe(bipy)_2(NCS)_2]$ at 298 and ~ 100 K from FIR measurements using the $^{54}Fe/^{57}Fe$ metal isotope technique

Compound	$\nu(^{54}Fe)^a$			
	at 298 K	at $\sim$ 100 K	Assignment	Ref.
$[Fe(phen)_2(NCS)_2]$ (I)	252.0 (s, br)	532.6 (m)^b 528.5 (m)	Fe-N(NCS)	33
	220.0 (m)	379.0 (m) 371.0 (w)	Fe-N(phen)	
$[Fe(phen)_2(NCSe)_2]$ (II)	228.0 (s, br)	530.5 (s) 527.0 (m)	Fe-N(NCSe)	77
	218 (sh)	366.0 (w) 360.0 (w)	Fe-N(phen)	
$[Fe(bipy)_2(NCS)_2]$ (III)	253.0 (s, br)	498.3 (s) 492.0 (w)	Fe-N(NCS)	77
	235 (sh)	393.0 (w) 374.7 (m)	Fe-N(bipy)	

^{a} The isotope effect $\Delta\bar{\nu} = \bar{\nu}(^{54}Fe\text{-}^{57}Fe)$ ranges from ~ 1 to ~ 6 cm^{-1}
^{b} The low-temperature Fe-N(ligand) stretching band is split into a doublet

Takemoto and Hutchinson have investigated the FIR spectra (600–100 cm^{-1}) at 298 and ~ 100 K of I [33] and of II and III [77]. They employed the metal isotope technique ($^{57}Fe/^{54}Fe$) for assigning the metal-ligand stretching vibrations. The observed isotope sensitive Fe–N(ligand) bands are listed in Table 1.

The large shift to higher frequencies on cooling indicates an enormous strengthening of all Fe–N(ligand) bonds. This is supported by X-ray data of König and Watson [76] on the spin crossover system III. They found that the bond lengths of Fe–N(NCS) shortened by 0.08 Å and those of Fe–N(bipy) shortened by 0.12 Å and 0.16 Å on cooling from 295 to 100 K.

Takemoto et al. have also measured the room temperature FIR spectra (600–140 cm^{-1}) of $[Fe(phen)_2X_2]$ complexes with X = Cl$^-$, Br$^-$, N$_3^-$, NCO$^-$, OAc$^-$, CNO$^-$, and CN$^-$ [50], again employing the $^{57}Fe/^{54}Fe$ metal isotope technique. The metal-ligand stretching vibrations have been found to depend strongly on the spin state of iron(II). For the high-spin complexes (X = Cl$^-$, Br$^-$, N$_3^-$, NCO$^-$, OAc$^-$) the Fe–N(phen) stretches were observed at ~ 225 cm^{-1} and the Fe–X stretches in the region of 325–180 cm^{-1}. For the low-spin complexes (X = CN$^-$, CNO$^-$) the Fe–N(phen) stretches were found at 350–395 cm^{-1} and the Fe–X stretches at 520–540 cm^{-1}. For comparison, the Fe–CN stretch in K$_4$[Fe(CN)$_6$] appeared at 588 cm^{-1}. Spin state dependent shifts of the ligand bands were also observed.

Ferraro and Takemoto [68] have used FIR spectroscopy to follow the extent of spin state conversion under pressure in I, II, and III. In good agreement with Fisher and Drickamer [78], who used Mössbauer spectroscopy to determine the relative changes of the fractions of the HS and the LS state in I and II and in other $[Fe(phen)_2X_2]$ complexes under applied pressure, Ferraro and Takemoto observed

Fig. 6. [57]Fe Mössbauer absorption spectra of [Fe(phen)$_2$(NCS)$_2$] as a function of   ▷
temperature (P. Ganguli, P. Gütlich, W. Müller, unpublished)

only partial conversion from HS to LS, which was reversible after release of pressure. Complete conversion with pressure to LS did not occur. This was possible, however, by cooling a pressurized sample of a mixture of both spin isomers to ~ 100 K in case of III. Heating the mixture to 353 K caused complete conversion to the HS isomer. Similar results were obtained with I and II. In view of these effects Ferraro and Takemoto predicted that, using a combination of pressure and temperature, it should be possible to completely convert the HS complexes [Fe(phen)$_2$X$_2$] (X = Cl$^-$, Br$^-$, N$_3^-$, and NCO$^-$) to the LS form, for which Fisher and Drickamer have observed only partial conversion using pressure alone [78].

A careful variable-temperature IR and FIR study was carried out on I and II by Sorai and Seki [34]. Particularly in the vicinity of the transition temperature T_c they recorded spectra at rather small temperature intervals. In accordance with the earlier report of Baker and Long [58], the C–N stretching modes of CNS$^-$ and CNSe$^-$ gave rise to two spin state dependent strong doublets, one at 2074/2062 cm^{-1} due to $\tilde{\nu}_1$(^{12}C–N), accompanied by a weaker doublet at 2040/2020 cm^{-1} due to $\tilde{\nu}_1$(^{13}C–N), in the ^{5}T$_2$(O$_h$) state, and the other at 2114/2107 cm^{-1} due to $\tilde{\nu}_1$(^{12}C–N), with a weaker doublet at 2081/2064 cm^{-1} due to $\tilde{\nu}_1$(^{13}C–N), in the ^{1}A$_1$(O$_h$) state. The simultaneous presence of these bands with temperature dependent relative intensities proved once more the coexistence of the HS and the LS ground states around T_c. Another important result of these studies is that the $\tilde{\nu}_1$(C–N) frequency itself in each spin state did not show any temperature dependence within the experimental error of 0.5 cm^{-1}. This implies that the bond strength and thus the bond length of the Fe–N(ligand) bonds in a particular spin state form may be considered approximately independent of temperature. This, in turn, lends support to the conclusion that, unlike previous predictions [11,79], the energy separation ϵ between the ^{1}A$_1$(O$_h$) ground state and the higher ^{5}T$_2$(O$_h$) state is nearly independent of temperature. For ϵ to vary with temperature one would expect the $\tilde{\nu}_1$(C–N) frequency to depend on temperature, too. Further evidence for this conclusion about ϵ could be obtained by Sorai and Seki from temperature dependent FIR spectra (400–30 cm^{-1}). In the vicinity of T_c the Fe–N bands arising from both ^{5}T$_2$ and ^{1}A$_1$ ground states coexisted well separated from each other, in accordance with earlier measurements [33], but without any appreciable temperature dependence of their frequencies.

3.1.3 [57]Fe Mössbauer Effect

König and Madeja [62,64] have first communicated results from [57]Fe Mössbauer effect measurements on I and II at 293 and 77 K. At both temperatures only one doublet was observed, but the quadrupole splitting ΔE_Q as well as the isomer shift changed drastically on going from 293 to 77 K in both systems: ΔE_Q ~ 2.6 mms^{-1} and δ ~ 1.0 mms^{-1} at 293 K, which is typical of high-spin iron(II) [80,81]; ΔE_Q ~

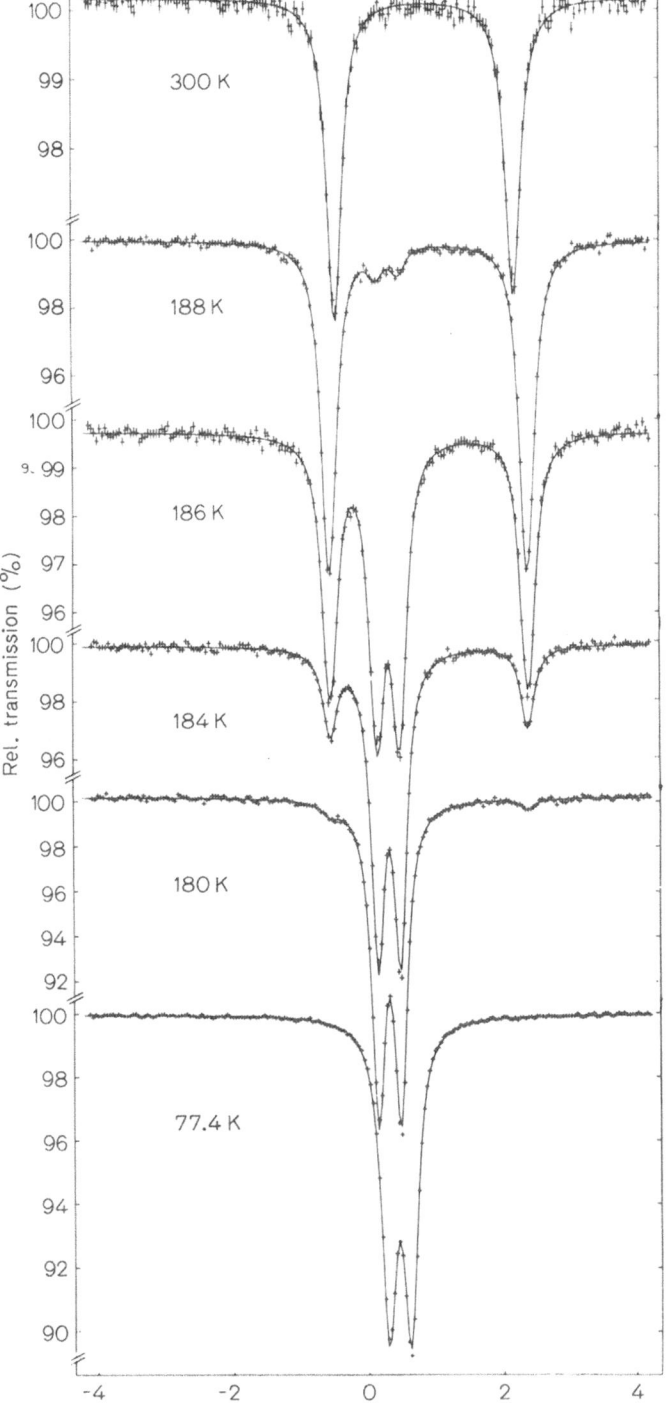

Fig. 7. ^{57}Fe Mössbauer spectra of [Fe$_x$Mn$_{1-x}$(phen)$_2$(NCS)$_2$] at 5 K as a function   ▷
of the iron concentration (from Ref. 179)

0.2–0.3 mms^{-1} and δ $\sim$ 0.35 mms^{-1} at 77 K, being typical data for a low-spin
iron(II) species.

A series of temperature dependent ^{57}Fe Mössbauer spectra of [Fe(phen)$_2$(NCS)$_2$]
between room temperature and 83 K, together with variable-temperature proton NMR
data, has first been presented by Dézsi et al.[63]. The Mössbauer spectra show two well
resolved quadrupole doublets with variable intensity as a function of temperature:
the doublet with the large quadrupole splitting (two outermost lines) dominates at
high temperatures and originates from the $^5T_2(O_h)$ state, the doublet with the small
quadrupole splitting (two inner lines) dominates at low temperatures and arises from
the $^1A_1(O_h)$ state. Some variable-temperature ^{57}Fe Mössbauer spectra of [Fe(phen)$_2$
(NCS)$_2$], which we have recently recorded at temperatures down to 4.2 K[70], are
shown in Fig. 6. The Mössbauer spectra represent another proof for the coexistence
of two spin isomers, similar to the variable-temperature IR spectra[34,58]. Furthermore, as the quadrupolar precession time of the $I = \frac{3}{2}$ state of ^{57}Fe is of the order of
10^{-7} s, which determines the time scale of ^{57}Fe Mössbauer spectroscopy to $\gtrsim 10^{-7}$ s,
the coexisting spin isomers must have lifetimes on the order of, or longer than,
10^{-7} s.

From variable-temperature ^{57}Fe Mössbauer spectra König and Ritter[82] determined the Debye-Waller factors f over the temperature range of 298–4.2 K in III (on
polymorph "one" with $\sim$ 0.9 B.M. at 77 K). Whereas the magnetic moment and
other physical properties show an abrupt change with temperature near T$_c$ ($\sim$ 213 K),
they found a continuous change of -lnf.

With the aim of studying the nature of the electronic ground state in I and in
[Fe(2-Cl-phen)$_3$] (ClO$_4$)$_2$, Reiff and Long[69] have placed the polycrystalline material
in a large external magnetic field and measured the ^{57}Fe Mössbauer spectrum at
300 K in case of I, and at 4.2 K in case of [Fe(2-Cl-phen)$_3$](ClO$_4$)$_2$. From the
magnetically perturbed quadrupole split spectra[83] the principal component V$_{zz}$
of the electric field gradient tensor turned out to be negative in both experiments.
This result was interpreted in terms of one electron occupying the d$_{z^2}$ orbital corresponding to an orbital singlet ground state (^5A) in I at 300 K (and well above T$_c$)
and in [Fe(2-Cl-phen)$_3$](ClO$_4$)$_2$ between 300 and 4.2 K.

In the course of an extensive investigation of the influence of metal dilution on
the spin crossover behaviour in iron (II) complexes[84-86], with the specific goal of
testing the cooperative domain model suggested by Sorai and Seki[34], we have explored the variable-temperature Mössbauer spectra of solid solutions of [Fe$_x$M$_{1-x}$
(phen)$_2$(NCS)$_2$] (M = Mn, Co, Ni, Zn)[15,179]. A few representative Mössbauer
spectra from our studies on [Fe$_x$Mn$_{1-x}$(phen)$_2$(NCS)$_2$] are displayed in Fig. 7. These
spectra demonstrate a general effect of metal dilution in iron(II) spin crossover
systems[15,84-86,179], viz. an increase of the relative stability of the HS state with
decreasing iron concentration. This is more explicitly reflected in the spin conversion curves x$_{HS}$(T) of the solid solutions of [Fe$_x$M$_{1-x}$(phen)$_2$(NCS)$_2$] (M = Mn, Co,

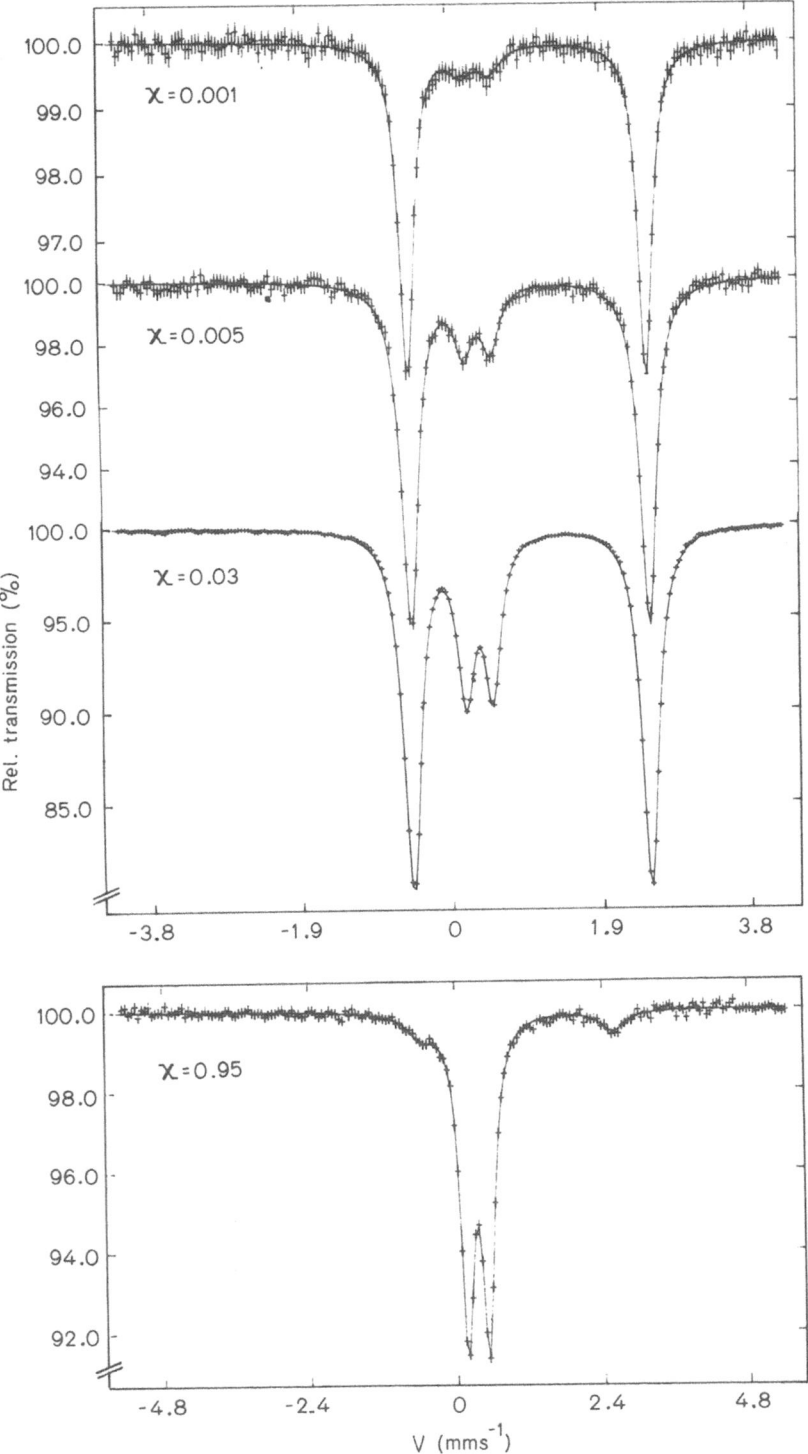

Fig. 8a–c. Temperature dependence of the area fraction x_{HS} of the HS quadrupole doublet of iron(II) in the solid solutions of (a) $[Fe_xMn_{1-x}(phen)_2(NCS)_2]$; (b) $[Fe_xCo_{1-x}(phen)_2(NCS)_2]$; (c) $[Fe_xZn_{1-x}(phen)_2(NCS)_2]$ (from Refs. 15, 179)

Zn) with variable iron concentration (Fig. 8a–c). This drastic change in the spin transition characteristics upon metal dilution appears to confirm the cooperative domain model of Sorai and Seki [34], who suggest that there exists a cooperative inter-action between the electronic state of the iron ion and the surrounding lattice. Furthermore, from an inspection of the spin conversion curves $x_{HS}(T)$, as shown in Fig. 8a–c it seems very likely that the residual HS fraction at the low temperature end and the residual LS fraction at the high temperature end is correlated with the ionic radius $r(M^{2+})$ of the transition metal ion of the host lattice. The relevant ionic radii are known to follow the order: $Mn^{2+}(0.82\ \text{Å}) > Fe^{2+}(0.74\ \text{Å}) \approx Zn^{2+} > Co^{2+}\ (0.72\ \text{Å}) > Ni^{2+}\ (0.70\ \text{Å})$. It is concluded from the Mössbauer studies [179] that the host lattice with $r(M^{2+}) > r(Fe^{2+})$ favours the stabilization of the HS state of iron(II) as in the case of $[Fe_xMn_{1-x}(phen)_2(NCS)_2]$, and that the host lattice with $r(M^{2+}) < r(Fe^{2+})$ favours the stabilization of the LS state as in the mixed crystals with M = Co, Ni. This ionic size effect may be rationalized in terms of an alteration of the Fe–N bond length with a concomitant change in the crystal field strength at the iron site: In the case of $r(M^{2+}) > r(Fe^{2+})$, a kind of "negative" lattice pressure is acting on the $[FeN_6]$ chromophore causing the Fe–N bond to increase and the crystal field strength to decrease, which increases the tendency to stabilize the HS state; in the case of $r(M^{2+}) < r(Fe^{2+})$, a "positive" lattice pressure acting on the $[FeN_6]$ chromophore forces the F–N bond to decrease and the crystal field strength to increase, which tends to favour the LS state.

A thorough reinvestigation [70] of the pure spin crossover system I, prepared by two different methods (A: extraction of one phen from $[Fe(phen)_3]\ (NCS)_2$ with acetone; B: adding dropwise a methanol solution of phen to a methanol solution containing Fe^{++} and NCS^-) have yielded some new aspects:

(i) The residual paramagnetism of a sample prepared by method (B) was found to be approximately 12% both from susceptibility *and* Mössbauer measurements, whereas none was found in sample A.

(ii) In the case of sample (B), as seen from Fig. 9, a marked change in the Mössbauer line width (Γ_{HS}), and a pronounced irregularity in the quadrupole splitting, $\Delta E_Q(HS)$, of the high-spin fraction in the temperature range 140–200 K ac-company the spin transition.

This has been considered to be indicative of a structural change accompanying the spin transition, possibly induced by a rotational order-disorder transition of the NCS^- ligand. The results from temperature dependent X-ray diffraction measurements on the polycrystalline samples A and B of I are in favour of a crystallographic phase change going along with the spin transition [179].

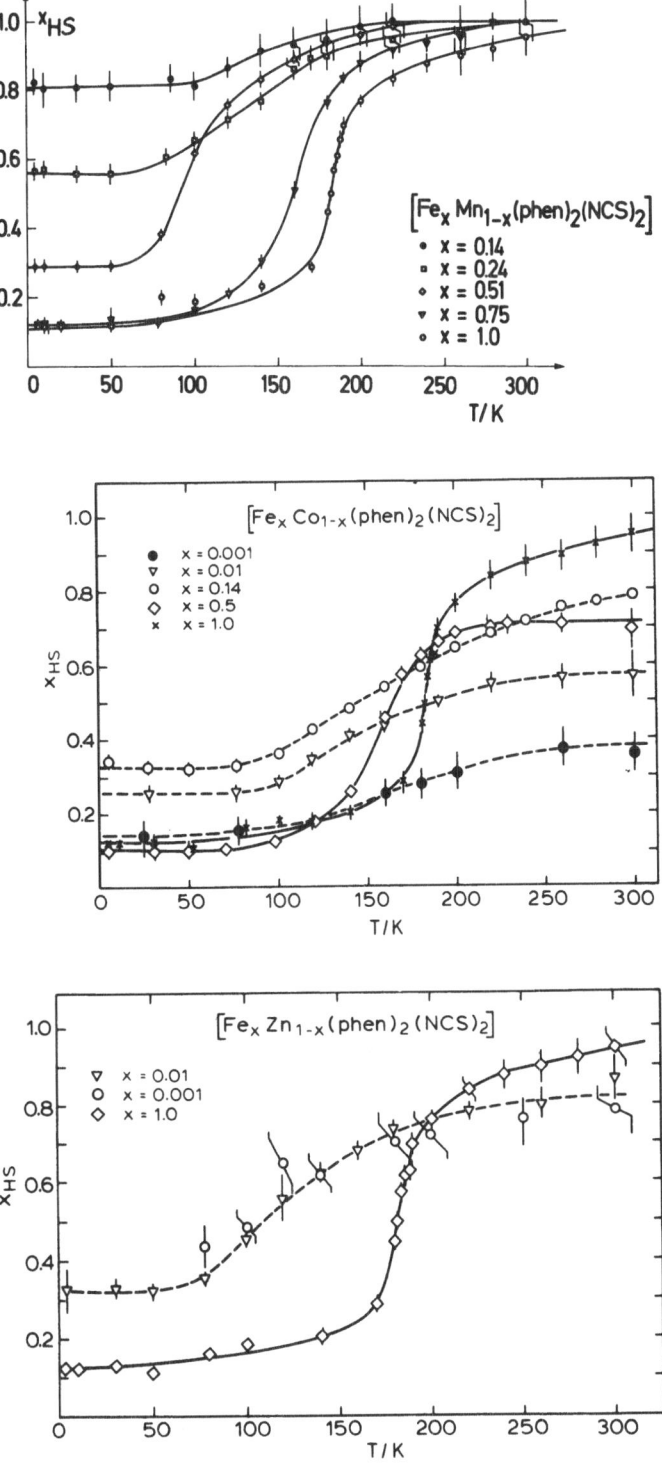

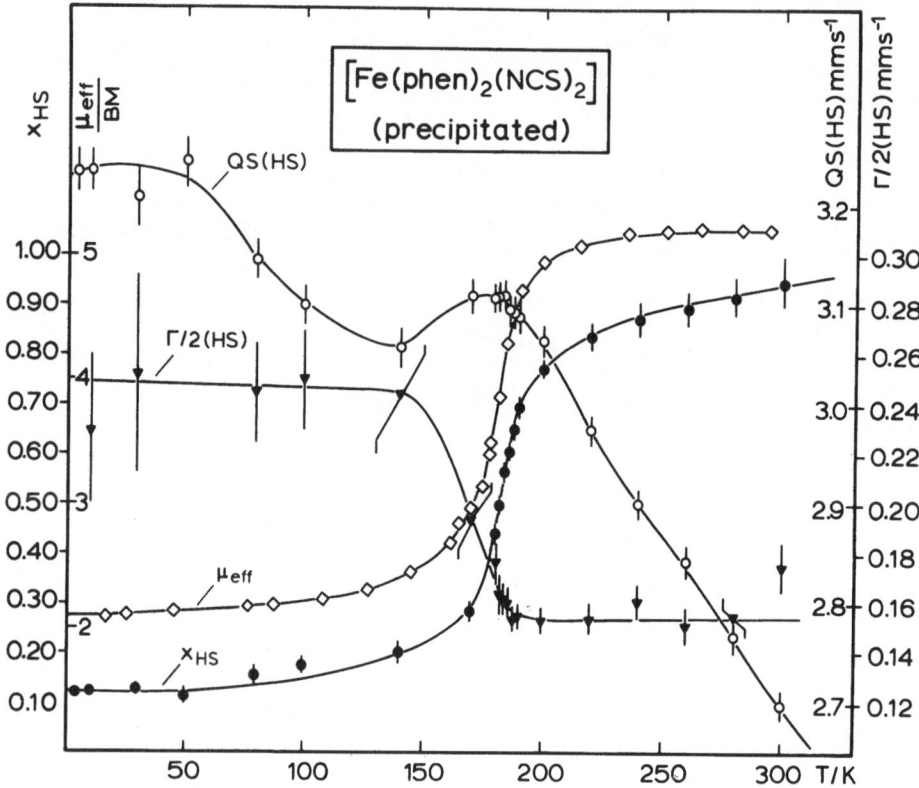

Fig. 9. Temperature dependence of the area fraction x_{HS}, the line width Γ_{HS}, and the quadrupole splitting $\Delta E_Q(HS)$ of the high-spin ($^5T_2$) quadrupole doublet, and of the magnetic moment μ_{eff} of [Fe(phen)$_2$(NCS)$_2$] (prepared after method (B); see text)[70]

3.1.4 Electronic Spectra

Electronic spectra (UV-vis) of I and II were measured at 298 and 77 K by König and Madeja[64]. The two compounds showed very similar absorption spectra at both temperatures. In the region characteristic of d—d transitions, the room-temperature spectra showed a band at 11900 cm^{-1} due to the $^5T_2(O_h) \rightarrow {}^5E(O_h)$ transition, whereas the 80 K spectra showed a band at 10400 cm^{-1}, which was assigned to the $^1A_1(O_h) \rightarrow {}^3T_1(O_h)$ transition. From the d—d transitions the following spectral parameters for both systems (I and II) have been estimated: $\Delta = 11900$ cm^{-1} and $B \sim 640$ cm^{-1} for the $^5T_2(O_h)$ state, $\Delta \sim 16300$ cm^{-1} and $B \sim 580$ cm^{-1} for the $^1A_1(O_h)$ state.

3.1.5 X-Ray Diffraction

Baker and Bobonich[5] have taken X-ray powder photographs of I and II at room temperature and below T_c. They found no structural change on cooling and thus ruled out a structural change as a possible explanation for the unusual magnetic behaviour. Essentially the same result has been obtained by König and Madeja[62,64].

In a recent study of the peak profiles of X-ray powder diffraction on [Fe(phen)$_2$(NCS)$_2$] as a function of temperature, particularly near the transition temperature T_c, different patterns have been found for the two spin state phases[70]. From these observations a structural change going along with the spin state change in I can no longer be ruled out.

Single-crystal X-ray analysis has been performed with the polymorphs "one" and "two" of III at room-temperature[71], and with polymorph "two" at 295 and 100 K[76]. Essential room-temperature data[71] are: orthorhombic with a = 16.04, b = 16.98, c = 15.94 Å, Z = 8, D_c = 1.482 g cm^{-3}, and space group Pbca (= D_{2h}^{15}) for polymorph "one"; orthorhombic with a = 13.17, b = 16.50, c = 10.08 Å, Z = 4, D_c = 1.469 g cm^{-3}, and space group Pcnb (= D_{2h}^{14}) for polymorph "two". Low-temperature ($\sim$ 100 K) measurements on III showed that there is no marked crystallographic change at the transition temperature. The decrease observed for all Fe–N bond lengths on going from 295 K to 100 K (the average changes are Δr = 0.08 Å for the Fe–N(NCS) bond and Δr = 0.14 Å for the Fe–N(phen) bond) reflects the strengthening of these bonds associated with the spin change and is consistent with results from FIR measurements[77]. The reported increase of the N–C bond length of the NCS$^-$ ligand[76] on passing from 295 K ($^5T_2$ state) to $\sim$ 100 K ($^1A_1$ state), however, disagrees with the observed shift of the C–N stretching frequency by some 40 cm^{-1} to higher frequencies on cooling[58,64,66], which indicates a strengthening of the C–N bond. Inspection of the bond angles shows that the octahedron of the FeN$_6$ chromophore is less distorted at $\sim$ 100 K than at room-temperature[76].

3.1.6 Heat Capacity Measurements

Sorai and Seki[34,87] realized that the spin transition in the present systems should be cooperative in nature, because of the abrupt change of the magnetic moment and of other molecular properties at a definite critical temperature T_c. They further realized that all the abundant experimental results accumulated thus far from magnetic susceptibility measurements and various other spectroscopic methods could only reveal changes in electronic and molecular structure associated with the spin transition, but could not allow any insight into possible long-range correlations as a characteristic feature of a cooperative spin transition. In order to proof the cooperative nature of the spin transition in I and II, Sorai and Seki performed precise heat capacity measurements between 375 and 13 K and determined the exact positions of the transition temperatures and the changes in enthalpy and entropy, ΔH and ΔS, associated with the spin transition. Fig. 10 shows the variation of the molar heat capacity C_p with temperature for II. C_p rises sharply at T_c = 231.26 K in case of II, and similarly at T_c = 176.29 K in case of I. A heat capacity jump ΔC_p (normal) = $C_p(HS)_{T_c} - C_p(LS)_{T_c}$ was observed at T_c for both compounds (cf. Table 2), from which the authors conclude that excitation of phonons is much easier in the high-spin (HS) phase than in the low-spin (LS) phase. This conclusion is based on the authors proposition that the normal heat capacity of each spin phase be composed of three significant parts,

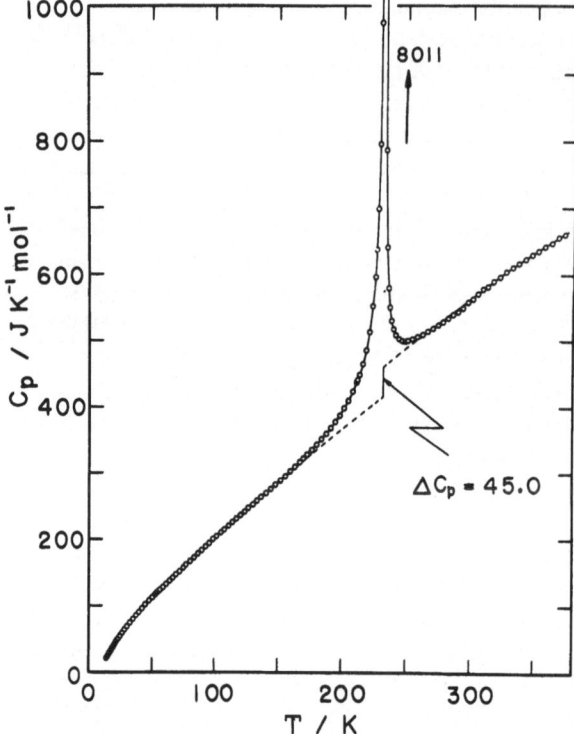

Fig. 10. Variation of the molar heat capacity C_p with temperature for [Fe(phen)$_2$(NCSe)$_2$]. Broken lines indicate the normal heat capacities (from Ref. 34)

$$C_p(\text{normal}) = C_{p,\text{latvib}} + C_{p,\text{intravib}} + C_{p,\text{mag}}. \tag{3.1}$$

$C_{p,\text{latvib}}$ is the contribution from lattice vibrations, $C_{p,\text{intravib}}$ the contribution from intramolecular vibrations, and $C_{p,\text{mag}}$ is the magnetic or electronic heat capacity arising from thermal excitation of electrons.

The gross changes in enthalpy (ΔH) and entropy (ΔS) associated with the spin transition were determined by Sorai and Seki as the excess part over the normal heat capacities; numerical values are summarized in Table 2. One result of utmost importance in understanding the driving force and eventually the mechanism of spin transition in the present as well as in other spin crossover systems is the fact that the total entropy change on going from the LS to the HS phase mainly arises from the

Table 2. Thermodynamic quantities associated with the spin transition in [Fe(phen)$_2$(NCS)$_2$] and [Fe(phen)$_2$(NCSe)$_2$] (from [34])

Compound	$\dfrac{T_c}{K}$	$\dfrac{\Delta H}{kJ\,mol^{-1}}$	$\dfrac{\Delta S}{J\,K^{-1}\,mol^{-1}}$	$\dfrac{\Delta C_p\ (\text{normal})}{J\,K^{-1}\,mol^{-1}}$
[Fe(phen)$_2$(NCS)$_2$]	176.29	8.60 ± 0.14	48.78 ± 0.71	18.7
[Fe(phen)$_2$(NCSe)$_2$]	231.26	11.60 ± 0.44	51.22 ± 2.33	45.0

vibrational contributions. The magnetic (electronic) contribution to ΔS is only about 27%, viz. $\Delta S_{mag} = R[\ln(2S+1)_{HS} - \ln(2S+1)_{LS}] = 13.38 \text{ J K}^{-1}$ as the contribution from the change in spin state ($S = 0 \rightarrow S = 2$), with the orbital contribution being zero for the present local symmetry of C_{2v}. About 50% of the remaining entropy change could be accounted for by changes in the Fe–N stretching frequencies; the other 50% are thought to be largely taken up by changes in N–Fe–N deformation vibrations, and a relatively small contribution is believed to arise from changes in the lattice vibrations. At any rate, from the large vibrational contribution to ΔS Sorai and Seki were led to conclude that "long-range correlation characteristic of the cooperative transition is established through the coupling between the electronic state and the phonon system".

3.1.7 Magnetic Resonance Studies

Comparatively few magnetic resonance experiments have been done so far on the present spin crossover systems.

In a brief communication Dézsi et al.[63] report on the unusual temperature dependence of the width ΔH/Gauss of the proton magnetic resonance signal in I (prepared by adding a pyridine solution of phen to a pyridine solution of [Fe(py)$_4$-(NCS)$_2$][48]). They observed a continuous rise in ΔH/Gauss with decreasing temperature followed by a sudden steep fall near T_c.

Paramagnetic resonance between 293 and 77 K on polycrystalline samples of III (polymorphs "one" and "two")[71] gave only an extremely weak signal at g ∼ 1.97. The signal intensity remains nearly constant on cooling to 77 K, whereas normally the intensity varies with $1/T$. This apparently is due to the fact that the expected increase in intensity is counterbalanced by the decrease in the number of paramagnetic ferrous ions.

ESR studies have recently been done on [Fe(phen)$_2$(NCS)$_2$] doped with 1 mol-% Mn^{2+}[207]. The spin Hamiltonian parameters D and E were found to decrease steadily with increasing temperature, but no change associated with the HS $\rightleftharpoons$ LS transition in the iron complex could be found.

3.2 [FeL$_2$X$_2$] Complexes with Substituted Phenanthroline and Substituted Bipyridine Derivatives (L = Y-phen, Y-bipy)

The substitution of one or more hydrogen atoms of the phen ligand by different substituents – alkyl, phenyl, halogen, NO$_2$ etc. – may alter the electronic structure and thus the magnetism of the iron (II) ion markedly. Two major factors may be operative to various extents depending on the nature of the substituent and the position in the phen molecule:
(i) electronic effects, i.e. electron donating and withdrawing power, respectively;
(ii) steric hindrance.
     This effect of intraligand substitution on the magnetism has been studied in many instances, not only in phenanthroline and bipyridine complexes.

Table 3. Magnetic data of bis(substituted-phenanthroline) complexes, [Fe(Y-phen)$_2$X$_2$]

Compound	T (K)	μ_{eff} (B.M.)	Ground state [a]	Ref.
[Fe(2−CH$_3$-phen)$_2$Cl$_2$]	293−77	5.30−5.15	$^5T_2$	88
[Fe(2−CH$_3$-phen)$_2$Br$_2$]	293−77	5.32−5.24	$^5T_2$	88
[Fe(2−CH$_3$-phen)$_2$(NCS)$_2$]	293−77	5.42−5.37	$^5T_2$	88
[Fe(2−CH$_3$-phen)$_2$(N$_3$)$_2$]·H$_2$O	293−77	5.24−5.07	$^5T_2$	88
[Fe(2−CH$_3$-phen)$_2$(N$_3$)$_2$]	293	5.09	$^5T_2$	88
[Fe(2−CH$_3$-phen)$_2$mal]·5H$_2$O	293−77	5.41−5.23	$^5T_2$	88
[Fe(2−CH$_3$-phen)$_2$mal]	293	5.22	$^5T_2$	88
[Fe(2−CH$_3$-phen)$_2$(CN)$_2$]	293−77	1.02−0.96	$^1A_1$	88
[Fe(4−CH$_3$-phen)$_2$(NCS)$_2$]	293−79	5.41−1.39	$^5T_2 \rightleftharpoons {}^1A_1$	90
[Fe(4−Cl-phen)$_2$(NCS)$_2$]	293−79.7	5.18−2.75	$^5T_2 \rightleftharpoons {}^1A_1$	90
[Fe(4−CN-phen)$_2$(NCS)$_2$]·$\frac{1}{2}$ pi	290−81	4.91−1.75	$^5T_2 \rightleftharpoons {}^1A_1$	90
[Fe(4−C$_2$H$_5$COO-phen)$_2$(NCS)$_2$]	374−81	5.38−1.03	$^5T_2 \rightleftharpoons {}^1A_1$	90
[Fe(4−C$_2$H$_5$COO-phen)$_2$(NCS)$_2$]·$\frac{1}{2}$ pi	333−78	4.76−0.70	$^5T_2 \rightleftharpoons {}^1A_1$	90
[Fe(4−C$_4$H$_9$COO-phen)$_2$(NCS)$_2$]	393/300/240/134/81.2	3.27/2.93/1.30/0.78/0.75	$^5T_2 \rightleftharpoons {}^1A_1$	93
[Fe(4,7−(C$_4$H$_9$COO)$_2$-phen)$_2$(NCS)$_2$]	376/290/185/86.9	2.81/1.97/1.45/1.38	$^5T_2 \rightleftharpoons ({}^3T_1, {}^1A_1)$	93
[Fe(4,7−(C$_6$H$_5$)$_2$-phen)$_2$(NCS)$_2$]	352/204/79	4.28/4.23/4.13	$^5T_2 + {}^3T_1$	93
[Fe(5−Cl-phen)$_2$(NCS)$_2$]·2H$_2$O	293−93	2.87−2.54	$^1A_1$	67
[Fe(5−NO$_2$-phen)$_2$(NCS)$_2$]·2H$_2$O	293−93	2.46−2.23	$^1A_1$	67
[Fe(5−CH$_3$-phen)$_2$(NCS)$_2$]·2H$_2$O	295−93	2.53−2.25	$^1A_1$	67
[Fe(5−Cl-phen)$_2$Cl$_2$]·2H$_2$O	320−90	5.31−5.30	$^5T_2$	67
[Fe(5−Cl-phen)$_2$(NCS)$_2$]·H$_2$O (preparation method 1)	296−93	5.04−3.55	$^5T_2 \rightleftharpoons {}^1A_1$	67
[Fe(5−Cl-phen)$_2$(NCS)$_2$]·H$_2$O (preparation method 2)	373−84	5.18−2.04	$^5T_2 \rightleftharpoons {}^1A_1$	67
[Fe(5−CH$_3$-phen)$_2$(NCS)$_2$]·H$_2$O (preparation method 1)	298−88	4.99−2.13	$^5T_2 \rightleftharpoons {}^1A_1$	67
[Fe(5−NO$_2$-phen)$_2$(NCS)$_2$]·H$_2$O (preparation method 1)	295−93	4.85−3.36	$^5T_2 \rightleftharpoons {}^1A_1$	67

Compound	T(K)	μ_{eff} (B.M.)	Ground state [a]	Ref.
[Fe(5−NO$_2$-phen)$_2$(NCS)$_2$]·H$_2$O (preparation method 2)	320−89	4.96−3.17	$^5T_2 \rightleftharpoons {}^1A_1$	67
[Fe(5−Ph-phen)$_2$(NCS)$_2$]·H$_2$O	292−93	3.68−2.80	$^5T_2 \rightleftharpoons {}^1A_1$	67
[Fe(4,7-(CH$_3$)$_2$-phen)$_2$(NCS)$_2$]	290/77.7/4.2	5.23/1.11/0.79	$^5T_2 \rightleftharpoons {}^1A_1$	95
[Fe(4,7-(CH$_3$)$_2$-phen)$_2$(NCS)$_2$]·pi	290.4−84.9	5.48−1.10	$^5T_2 \rightleftharpoons {}^1A_1$	90
[Fe(4,7-(CH$_3$)$_2$-phen)$_2$(NCS)$_2$]·2py	293−80.5	5.49−5.25	$^5T_2$	90
[Fe(4,7-(C$_2$H$_5$COO)$_2$-phen)$_2$(NCS)$_2$]	383/189/79.8	2.91/2.79/2.71	$^3T_1$(?)	90
[Fe(4,7-Cl$_2$-phen)$_2$(NCS)$_2$]	285−80.4	5.12−5.02	$^5T_2 + {}^3A_2$[b]	90
[Fe(4,7-Cl$_2$-phen)$_2$(NCS)$_2$]·pi	293−78.7	5.17−5.03	$^5T_2 + {}^3A_2$[b]	90
[Fe(3,4,7,8-(CH$_3$)$_4$-phen)$_2$(NCS)$_2$]·2H$_2$O	357/293/83.4	5.41/5.39/5.48	$^5T_2$	67
[Fe(3,4,7,8-(CH$_3$)$_4$-phen)$_2$Cl$_2$]·2H$_2$O	289−95	5.40−5.37	$^5T_2$	67

[a] In the approximation of O_h symmetry
[b] Coexisting states suggested from two overlapping doublets in the Mössbauer spectra. $^3A_2$ is a crystal field state of $^3T_1$(O_h)

3.2.1 [Fe(Y-phen)₂X₂] Complexes with Mono-substituted Phenanthroline Derivatives

The compounds $[Fe(2\text{-}CH_3\text{-}phen)_2X_2]$ with $X = Cl^-$, Br^-, NCS^-, N_3^-, and 1/2 mal are all high-spin with a $^5T_2(O_h)$ ground state and an effective magnetic moment μ_{eff} at 293 K ranging from 5.09 to 5.42 B.M.[88]. The susceptibilities follow the Curie-Weiss law, $\chi_m = C_m/(T-\theta)$ between 293 and 77 K, with $\theta = -2$ to -6. The effective magnetic moments are listed in Table 3. The ^{57}Fe Mössbauer isomer shifts δ at room temperature are ~ 1.0 mms^{-1}; the quadrupole splitting ΔE_Q ranges between 1.84 and 2.58 mms^{-1} (at 294 K) and is temperature dependent[88]. These data are typical of iron (II) in the high-spin state. The electronic spectra show the $^5T_2(O_h) \rightarrow {}^5E(O_h)$ transition at ~ 10000 cm^{-1}, with the $^5E(O_h)$ state being split by 2000–3000 cm^{-1}[88].

The compound $[Fe(2\text{-}CH_3\text{-}phen)_2(CN)_2]$ is low-spin and has a $^1A_1(O_h)$ ground state, with $\mu_{eff} = 1.02$ B.M. at 293 K, and $\delta = 0.06$ mms^{-1}, $\Delta E_Q = 0.88$ mms^{-1} at 294 K[88].

Obviously, the ligand field strength in this family of bis(2-CH$_3$-phen) complexes is reduced as compared to their unsubstituted analogs, $[Fe(phen)_2X_2]$. This follows immediately from the magnetic moments of the compounds with $X = 1/2$ mal or NCS^-: $[Fe(phen)_2 mal] \cdot 7 H_2O$ has a spin-triplet ground state[56], whereas the analogous 2-CH$_3$-phen complex (with 5 molecules of crystal water though, but this difference in content of crystal water is not considered responsible for the reduction in field strength in the present case) has a spin quintet ground state. The spin crossover phenomenon in $[Fe(phen)_2(NCS)_2]$ is completely quenched in the 2-CH$_3$-phen analog. The other complexes with $X = Cl^-$, Br^-, and N_3^- have $^5T_2(O_h)$ ground states in both the substituted and the unsubstituted cases. The reduction of the field strength can also be seen in the electronic spectra: The d–d doublet arising from the $^5T_2 \rightarrow {}^5E$ transition (with ^5E being split by 2000–3000 cm^{-1}) appears at somewhat lower frequencies in the bis(2-CH$_3$-phen) complexes[88] as compared to the corresponding unsubstituted complexes[48]. As the difference between corresponding 10 Dq values is only on the order of a few hundred wavenumbers, one has suspected[88] that the crossover between the ground states $^5T_2$ and $^1A_1$ in the $[Fe(2\text{-}CH_3\text{-}phen)_2X_2]$ series lies very close to the position of the 2-CH$_3$-phen ligand in the spectrochemical series. This receives support from the magnetic behaviour of the tris(2-CH$_3$-phen) complex: the $[Fe(2\text{-}CH_3\text{-}phen)_3]^{2+}$ complex ion shows a temperature-dependent (rather gradual) spin transition[89], whereas the unsubstituted parent complex $[Fe(phen)_3]^{2+}$ is low-spin.

It is clear from these results that the effect of replacing a hydrogen atom by a methyl group in 2-position of phen is (at least) twofold, electronic and steric. CH$_3$ has a stronger σ-donating power than hydrogen, increasing the basicity of the nitrogen atom and making the Fe–N σ-bond to the 2-CH$_3$-phen ligand somewhat stronger than that to the phen ligand. This electronic effect by itself causes a partial increase of the ligand field strength, but is more than counterbalanced by the opposing steric effect. The CH$_3$ substituent requires considerably more space than a hydrogen atom. The consequence is that, in order to avoid too much steric interference with the neighbouring ligands, the Fe–N(2-CH$_3$-phen) bond length R will increase, which in turn causes a reduction in field strength according to $\Delta \sim R^{-5}$ [3]. As can be seen from the magnetic behaviour, the steric effect obviously dominates over the electronic effect in the [Fe(2-CH_3-phen)$_2$X$_2$] complexes.

The situation is different if a methyl group, or other substituents, is introduced in 4-position of phen. The model shows that a large substituent in the 4-position is far less able to interfere with neighbouring ligands of the same complex molecule, and therefore the steric effect on the magnetic behaviour of the central ion should play only a minor role, and the electronic effect — here the enhanced σ-donating power of CH_3 as compared to H, causing an overall increase of the ligand field strength as compared to the 2-CH_3-phen complex — should prevail. This is indeed the case.

1,10-phenanthroline (phen)

[Fe(4-CH_3-phen)$_2$(NCS)$_2$] shows a temperature dependent $^5T_2(O_h) \rightleftharpoons {}^1A_1(O_h)$ transition, less abrupt, though, than in case of the unsubstituted compound [Fe(phen)$_2$(NCS)$_2$][90]. The transition temperature in the 4-CH_3-phen complex is shifted somewhat to higher temperatures as compared to the unsubstituted complex, indicating a somewhat higher ligand field strength in the 4-CH_3-phen complex due to a slight increase in ligand → metal σ-donation on substituting hydrogen for a CH_3 group. Spin crossover has also been found in other [Fe(4-Y-phen)$_2$(NCS)$_2$] complexes, where Y = Cl, CN with 1/2 α-picoline as crystal solvent, C_2H_5COO (without and with 1/2 α-picoline)[90]; cf. Table 3. The shape and the position of the curve for the temperature dependence of μ_{eff} varies with the nature of the substituent (see Fig. 11).

The high-spin ($^5T_2$) $\rightleftharpoons$ low-spin ($^1A_1$) transition in [Fe(4-CH_3-phen)$_2$(NCS)$_2$] has been further persued by magnetic susceptibility measurements between 0.98 and 304 K, and by ^{57}Fe Mössbauer effect measurements between 4.2 and 309 K[91,92]. From the χ(T) data, a transition temperature has been found at 215 K. μ_{eff} decreases smoothly from 5.223 B.M. at 303.7 K to 1.417 B.M. at 77 K, levelling off down to about 20 K. Below 20 K there is a pronounced decrease of μ_{eff}, arising from the depopulation of all paramagnetic states contributing to the overall magnetic moment. The ^{57}Fe Mössbauer spectra show two well-resolved quadrupole doublets with strongly temperature-dependent intensities. Their isomer shift and quadrupole splitting data at 215 K are (in mms^{-1}) : δ = 0.99, ΔE_Q = 2.88 for the HS($^5T_2$) state; δ = 0.36, ΔE_Q = 0.37 for the LS ($^1A_1$) state (relative to natural ion at 298 K). Debye-Waller factors were determined for the two states. The values -lnf($^5T_2$) and -lnf($^1A_1$) follow the Debye model between 175 and 250 K with θ_D ($^5T_2$) = 126 K and between 105 and 225 K with θ_D ($^1A_1$) = 150 K, respectively. Deviations encountered outside these regions were considered as evidence for the formation of cooperative domains as suggested by Sorai and Seki[34,87]. The difference between the Debye temperatures θ_D ($^5T_2$) and θ_D ($^1A_1$) may well be understood in terms of more rigidity in the lattice of the $^1A_1$ state as compared to that of the $^5T_2$ state. A study of the magnetic hyperfine interaction at 4.2 K yielded $V_{zz}(^1A_1) < 0$. $V_{zz}(^5T_2) > 0$, however, was concluded from the spin reversal of the texture-induced asymmetry of the Mössbauer line intensities.

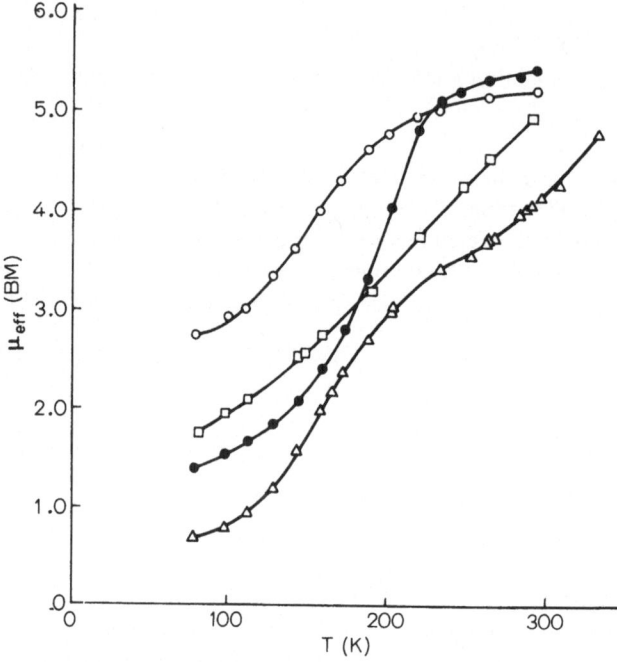

Fig. 11. Temperature dependence of the effective magnetic moment μ_{eff} in: [Fe(4-CH$_3$-phen)$_2$-(NCS)$_2$] (●); [Fe(4-Cl-phen)$_2$(NCS)$_2$] (○); [Fe(4-CN-phen)$_2$(NCS)$_2$]·$\frac{1}{2}$α-picoline (□), and [Fe(4-C$_2$H$_5$COO-phen)$_2$(NCS)$_2$]·$\frac{1}{2}$α-picoline (▽) (from Ref. 90)

Most interesting is a recent communication of Madeja et al.[93], who have introduced the bulky n-butyl ligand in 4-position of phen and studied the ground state properties of [Fe(4-C$_4$H$_9$-phen)$_2$(NCS)$_2$] by magnetic susceptibility (between 77 and 400 K) and Mössbauer measurements (between 93 and 298 K). μ_{eff} varies gradually from 0.75 B.M. at 81 K to 3.27 B.M. at 393 K, suggesting that, due to strong low-symmetry ligand field distortions rendering the $^3T_1$ state thermally accessible, a $^1A_1(O_h) \rightleftharpoons {}^3T_1(O_h)$ transition might be operative here. The room-temperature Mössbauer spectrum, however, with two separate doublets, one with $\delta = 0.96$ mms^{-1} (relative to stainless steel) and $\Delta E_Q = 3.65$ mms^{-1} being typical of a $^5T_2$ ground state, and the other with $\delta = 0.42$ mms^{-1} and $\Delta E_Q = 0.49$ mms^{-1} being typical of a $^1A_1$ ground state, calls for an incomplete $^1A_1(O_h) \rightleftharpoons {}^5T_2(O_h)$ transition.

The effect of intra-ligand substitution in the 5-position of phen has been studied on the family of [Fe(5-Y-phen)$_2$(NCS)$_2$]·nH$_2$O complexes by susceptibility and Mössbauer measurements[67]. These studies also revealed a distinct dependency of the magnetism on the amount of crystal water (see Table 3). The dihydrates with Y = 5-CH$_3$, 5-NO$_2$, and 5-Cl have essentially temperature-independent magnetic moments around 2.4–2.8 B.M. and also temperature-independent quadrupole splittings around 0.2–0.3 mms^{-1}; these data are characteristic of iron (II) low-spin compounds. The corresponding monohydrates show very smooth high spin $\rightleftharpoons$ low spin transition with

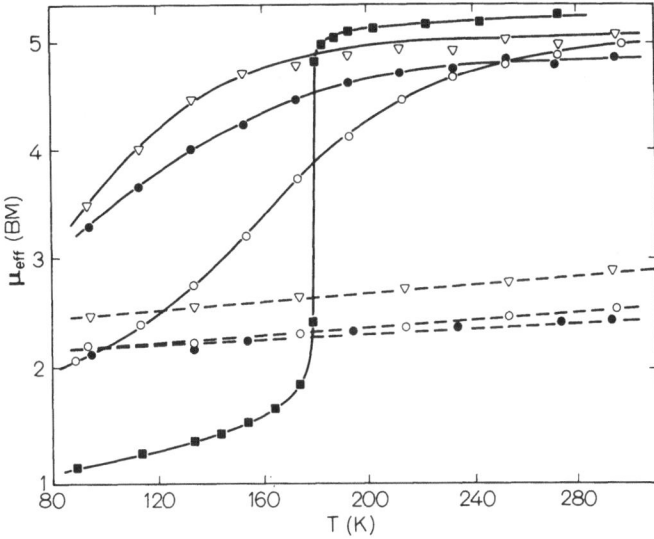

Fig. 12. Temperature dependence of the effective magnetic moment μ_{eff} in [Fe(5-Y-phen)$_2$-(NCS)$_2$]·nH$_2$O. Y = H (■), CH$_3$ (○), Cl (▽), and NO$_2$ (●), respectively; broken lines for dihydrates (n = 2), solid lines for monohydrates (n = 1) (from Ref. 67)

considerable differences in their μ(T) characteristics as arising mainly from the different electronic effects caused by the 5-Y substituents (see. Fig. 12). The monohydrate of the unsubstituted compound, [Fe(phen)$_2$(NCS)$_2$]·H$_2$O, as with the anhydrous complex, shows a very abrupt spin change near 180 K. The dihydrate, however, seems to be low-spin judging from the low magnetic moments of ca. 2.97 (298 K) − 2.80 (88 K) B.M. Similar influences of the content of crystal water on the magnetic behaviour has also been observed in [Fe(2-pic)$_3$] Cl$_2$·nH$_2$O [84] and in a number of other solvates of spin crossover compounds (vide infra). There is strong evidence, even from x-ray crystal structure analysis [31,94,180,181], that hydrogen-bonding plays a key-role in the spin transition mechanism in solid solvates.

It is interesting to note that [Fe(5-Cl-phen)$_2$Cl$_2$]·2 H$_2$O is high-spin over the entire temperature range studied, whereas the corresponding [Fe(5-Cl-phen)$_2$(NCS)$_2$]· 2 H$_2$O complex with NCS⁻ instead of Cl⁻ as the coordinated anion shows low-spin (^{1}A$_1$) behaviour. This difference is readily explained by the different positions of NCS⁻ and Cl⁻ in the spectrochemical series, i.e. NCS⁻ causes a higher partial cubic field strength than Cl⁻.

3.2.2 [Fe(Y-phen)$_2$X$_2$] Complexes with Bis- and Poly-substituted Phenanthroline Derivatives

[Fe(4,7-(CH$_3$)$_2$-phen)$_2$(NCS)$_2$], in which the H atoms in the 4,7-positions of phen are replaced by a CH$_3$ group each, shows high-spin ⇌ low-spin transition (see Table 3) with a very abrupt change of the magnetic moment at T$_c$ = 121.5 K [95], probably indicating a first-order phase change, similar to that in the unsubstituted parent system [Fe(phen)$_2$(NCS)$_2$] [34]. The transition temperature T_c, however, is in the present system about 55 K lower than in [Fe(phen)$_2$(NCS)$_2$]. The monosubstituted analog, [Fe(4-CH$_3$-phen)$_2$(NCS)$_2$], which also shows spin crossover [90,92], differs markedly from the unsubstituted as well as from the 4,7-(CH$_3$)$_2$-phen complexes in that its magnetic moment changes gradually with temperature.

The Mössbauer spectra of [Fe(4,7-(CH$_3$)$_2$-phen)$_2$(NCS)$_2$] exhibit two well resolved doublets due to the coexisting $^5T_2$(HS) and $^1A_1$(LS) states. From the saturation corrected areas of the Mössbauer spectra, Debye-Waller factors were determined between 4.2 and 298 K. At T_c, -lnf$_{total}$ shows a discontinuity of $\sim 20\%$ corresponding to a change between f($^5T_2$) and f($^1A_1$), which the authors interpret in terms of different lattice dynamical properties at the sites of $^5T_2$ and $^1A_1$ ground state iron(II) molecules in the crystal. A thorough reinvestigation of the Mössbauer spectra of [Fe(4,7-(CH$_3$)$_2$-phen)$_2$(NCS)$_2$], taken at short temperature intervals, revealed a hysteresis effect of ca. 3 K width, with $T_c^> = 121.7$ K for rising and $T_c^< = 118.6$ K for lowering of temperature [96]. These observations could only qualitatively be reproduced with the thermodynamic model of Slichter and Drickamer [97].

Distinct and different peak profiles of X-ray powder diffraction on [Fe(4,7-(CH$_3$)$_2$-phen)$_2$(NCS)$_2$] have been observed for the $^5T_2$ and $^1A_1$ phases, respectively [183]. The intensities of the lines show the same temperature dependence and the same hysteresis behaviour as the area fractions of the HS and the LS quadrupole doublets, respectively, of the Mössbauer spectra. These studies [183] provide clear evidence for a simultaneous change of the electronic spin state and of the crystallographic properties.

The picoline solvate of [Fe(4,7-(CH$_3$)$_2$phen)$_2$(NCS)$_2$]·pi also exhibits a sharp (discontinuous) spin transition at $T_c \sim 200$ K [90]. The same complex compound, crystallized with two pyridine molecules per unit formula though, is purely high-spin over the whole temperature range under study (see Table 3). This is another example of a pronounced influence arising from the amount and the nature of crystallizing solvent molecules. The compounds with Y = 4,7-(C$_5$H$_5$COO)$_2$, Cl$_2$ and X = NCS$^-$ could not unambiguously be characterized with respect to their ground state. A $^3A_2$ ground state has been considered likely in the former case, and a mixture of $^5T_2$ and $^3A_2$ ground states in the latter case, where Mössbauer spectroscopy reveals two separate spectra.

Madeja et al. [93] have studied the effect of intra-ligand substitution in similar complexes, though with very bulky substituents in 4,7 positions of phen: [Fe(4,7-(C$_4$H$_9$COO)$_2$-phen)$_2$(NCS)$_2$] has been found to behave similarly in magnetism as its monosubstituted analog with Y = 4-C$_4$H$_9$COO, X = NCS$^-$ [93]; the magnetic moments in the bis-substituted case range from 1.38 B.M. at 86.9 K to 2.81 B.M. at 376 K (see Table 3 and Fig. 13). From these data as well as from Mössbauer measurements it was not possible to clarify unambiguously the ground state properties. The authors suggest that at 93 K a mixture of spin-singlet and spin-triplet ground states is present, whereas a partial conversion into the $^5T_2$ state occurs at higher temperatures. A mixture of $^5T_2$(O$_h$) and $^3T_1$(O$_h$) ground states, but with no temperature-dependent spin transition, was suggested to exist in [Fe(4,7-(C$_6$H$_5$)$_2$-phen)$_2$(NCS)$_2$], which has magnetic moments around 4.2 B.M. essentially independent of temperature (see Table 3 and Fig. 13).

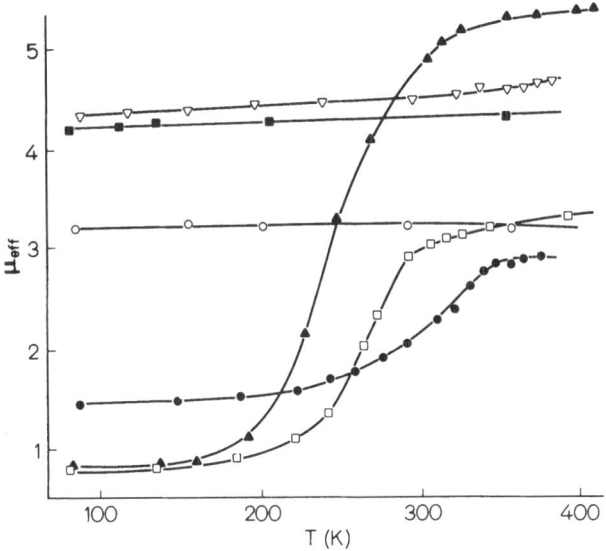

Fig. 13. Temperature dependence of the effective magnetic moment μ_{eff} of $[FeL_2(NCS)_2]$ with: L = 4-C_4H_9COO-phen (□); L = 4,7-$(C_4H_9COO)_2$-phen (●); L = 4,7-$(C_6H_5)_2$-phen (■); L = 4,4'-$(C_4H_9COO)_2$-bipy (▽); L = pip (▲); L = pmi (○) (from Ref. 93)

The $[Fe(Y\text{-}phen)_2(NCS)_2]$ complexes with Y = 2,9-$(CH_3)_2$, 5,6-$(CH_3)_2$ show high-spin behaviour, with the steric effect most probably prevailing in these cases. They partially convert into the low-spin form under pressure [98].

In the complexes $[Fe(3,4,7,8\text{-}(CH_3)_4\text{-}phen)_2X_2]$ with X = NCS^-, Cl^- one has observed pure high-spin behaviour [67]. Undoubtedly, the steric effect is expected here to dominate over the inductive effect.

3.2.3 $[Fe(Y\text{-}bipy)_2X_2]$ Complexes with Substituted Bipyridine Derivatives

Complex compounds of the type $[Fe(Y\text{-}bipy)_2X_2]$ have only rarely been investigated with respect to their magnetic behaviour. Both $[Fe(4,7\text{-}(CH_3)_2\text{-}bipy)(NCS)_2]\cdot H_2O$ and $[Fe(4,7\text{-}(CH_3)_2\text{-}bipy)_2(NCSe)_2]\cdot H_2O$ have been reported to show high-spin $\rightleftharpoons$ low-spin transition [67]. The magnetic moment changes gradually over a large temperature range as indicated by the following representative data: $\mu(T/K)/\mu_B$ = 0.88 (88.5), 0.89 (180), 2.90 (300), 4.59 (357), 5.24 (423), 5.40 (506) for the former complex, and 0.97 (79), 2.64 (300), 4.84 (381) for the latter complex [67].

Another example has been reported by Madeja et al. [93]. They observed the magnetic moment of $[Fe(4,7\text{-}(C_4H_9COO)_2\text{-}bipy)_2(NCS)_2]$ to vary only little with temperature with values between 4.69 B.M. (380 K) and 4.26 B.M. (85 K). The Mössbauer spectra at 93 and 293 K show two doublets each, with essentially no temperature variation of the intensities. From these experimental findings the authors conclude that a mixture of spin-quintet and spin-triplet ground states may be present in this compound.

3.3 Other [Fe(diimine)$_2$(NCS)$_2$] Complexes

Partial replacement of py in [Fe(py)$_4$(NCS)$_2$] by phen leads to a mixed-ligand complex containing three different kinds of N-coordinating ligands, [Fe(py)$_2$phen(NCS)$_2$] [99]. This complex compound undergoes abrupt spin transition at ca. 185 K, with magnetic moments of 5.10 B.M. at room temperature and 2.09 B.M. as the limiting value below T_c. A d–d transition band due to $^5T_2(O_h) \rightleftharpoons {}^5E(O_h)$ appears at 11 700 cm^{-1}. Mössbauer spectra recorded at 298 and 77 K [100] contain two separate doublets, the one at room temperature arises from the $^5T_2(O_h)$ state with $\delta = 0.91$ mms^{-1} (rel. to natural iron) and $\Delta E_Q = 2.46$ mms^{-1} and the 77 K spectrum essentially contains the $^1A_1(O_h)$ doublet with $\delta = 0.38$ mms^{-1} and $\Delta E_Q = 0.28$ mms^{-1} and some minor fraction of the $^5T_2$ doublet. The replacement of one phen ligand by two monodentate py ligands does not change significantly the crystal field properties and thus the electronic structure of the iron (II) ion. This follows immediately from a comparison of the magnetic behaviour and the Mössbauer and electronic spectral data of the related complexes [Fe(phen)$_2$(NCS)$_2$] and [Fe(py)$_2$phen(NCS)$_2$] [99,100]. Pronounced differences may only occur in the amount of the residual paramagnetism, depending on the method of sample preparation.

The replacement of one NCS$^-$ ligand in [Fe(bipy)$_2$(NCS)$_2$] by a py ligand, which is known to possess stronger π-back-bonding properties than NCS$^-$, will cause an increase in the ligand field strength.

The Mössbauer spectra of [Fe(bipy)$_2$py(NCS)], taken at 293 and 77 K, contain only one doublet each, with data typical of iron (II) low-spin [56]. The same behaviour was observed on [Fe(bipy)$_2$py(NCSe)] [56].

Madeja et al. [93] recently reported on the different magnetic behaviour of [Fe(pip)$_2$(NCS)$_2$] (pip = 2-pyridinaldehyde-N-isopropylimine) as compared to that of [Fe(pmi)$_2$(NCS)$_2$] (pmi = 2-pyridinaldehyde-N-methylimine). The former system shows a gradual $^5T_2(O_h) \rightleftharpoons {}^1A_1(O_h)$ spin transition (see Fig. 13) with magnetic moments from 5.31 B.M. at 408 K to 0.77 B.M. at 80 K. The latter complex has been suggested to contain a mixture of $^5T_2(O_h)$ and $^3T_1(O_h)$ ground states with no spin crossover; the magnetic moment remains practically constant around 3.15 B.M. This difference in magnetic properties most likely arises from a temperature dependent steric hindrance effect in case of the large isopropyl substituent as compared to the much smaller methyl group in the pmi complex.

4 Tris(substituted-phenanthroline) Complexes, [Fe(Y-phen)$_3$] X$_2$

The effect of intra-ligand substitution on the magnetism of tris(Y-phen) complexes of iron (II) has been demonstrated in many instances. The parent complex ion [Fe(phen)$_3$]$^{2+}$, possesses a $^1A_1(O_h)$ ground state at all temperatures independent of the nature of the non-coordinate anion. The substituted analogs with one or more hydrogen atoms of phen replaced by different substituents, may be classified into three groups according to their magnetic properties: (a) pure low-spin ($^1A_1$), (b) pure high-spin ($^5T_2$), and (c) $^5T_2(O_h) \rightleftharpoons {}^1A_1(O_h)$ crossover compounds. The magnetism depends on the number, the nature, and the position of the substituents. As discussed above, a combined action of electronic and steric effects must be considered responsible in determining the actual spin state.

4.1 [Fe(2-Y-phen)$_3$] X$_2$ Complexes

The most prominent spin crossover examples in this class of compounds are the [Fe(2-CH$_3$-phen)$_3$] X$_2$ complexes. As early as 1962, Irving and Mellor[101] noticed that [Fe(2-CH$_3$-phen)$_3$]$^{2+}$ salts are paramagnetic at room temperature. The stronger ligand-metal σ-interaction in the 2-CH$_3$-phen complex, which is caused by the higher inductive effect of a CH$_3$ group as compared to hydrogen, is more than outweighted by the steric hindrance, which tends to lengthen and thus weaken the L–M bond resulting in a stabilization of the $^5T_2(O_h)$ ground state.

Goodwin and Sylva[89] first investigated the magnetism of the salts [Fe(2-CH$_3$-phen)$_3$] X$_2$, X = I$^-$, ClO$_4^-$, BF$_4^-$, PF$_6^-$ between ~ 100 and 300 K, and found that in all cases the magnetic moment decreases smoothly on lowering the temperature (see Table 4). The shape of the $\mu_{eff}(T)$ curve shows a distinct dependency on the nature of the anion X. A later reinvestigation of the magnetic susceptibility down to ~ 2 K [102] revealed a discontinuity in the gradual decrease of $\mu_{eff}(T)$ around 70 K (see Fig. 14). Around this temperature the quadrupole splitting of the $^5T_2(O_h)$ state also shows irregularities in its $\Delta E_Q(T)$ behaviour, and we believe that these anomalies may be due to an hitherto unspecified phase transition. Goodwin and Sylva[89] concluded correctly that a thermal equilibrium between the nearly equi-energetic ground terms $^5T_2(O_h)$ and $^1A_1(O_h)$ is the cause for the anomalous behaviour of $\chi(T)$, going along with a colour change from orange (~ 300 K) to red-purple (~ 90 K), in all these

Table 4. Magnetic data of [Fe(2–Y-phen)$_3$]X$_2$ complexes

Compound	T (K)	μ_{eff} (B.M.)	Ground state	Ref.
[Fe(2–CH$_3$-phen)$_3$]X$_2$,				
X = I$^-$	295/155/91	5.06/4.62/4.00	$^5T_2 \rightleftharpoons {}^1A_1$	89
X = BF$_4^-$	303/170/97	5.48/4.88/3.41	$^5T_2 \rightleftharpoons {}^1A_1$	89
X = PF$_6^-$	296/162/92	5.30/4.92/3.91	$^5T_2 \rightleftharpoons {}^1A_1$	89
X = ClO$_4^-$	294/160/98	5.39/4.34/3.01	$^5T_2 \rightleftharpoons {}^1A_1$	89
X = ClO$_4^-$	300/200/150/ 100/50/2.6	5.38/5.07/4.11/ 2.58/2.30/1.48	$^5T_2 \rightleftharpoons {}^1A_1$	102
[Fe(2–CH$_3$O-phen)$_3$]- (ClO$_4$)$_2 \cdot$H$_2$O	300/200/100/ 20/2.60	5.31/3.81/1.73/ 1.37/0.87	$^5T_2 \rightleftharpoons {}^1A_1$	105
[Fe(2–Cl-phen)$_3$](ClO$_4$)$_2$	293/197/102	5.36/5.29/5.00	$^5T_2$	111

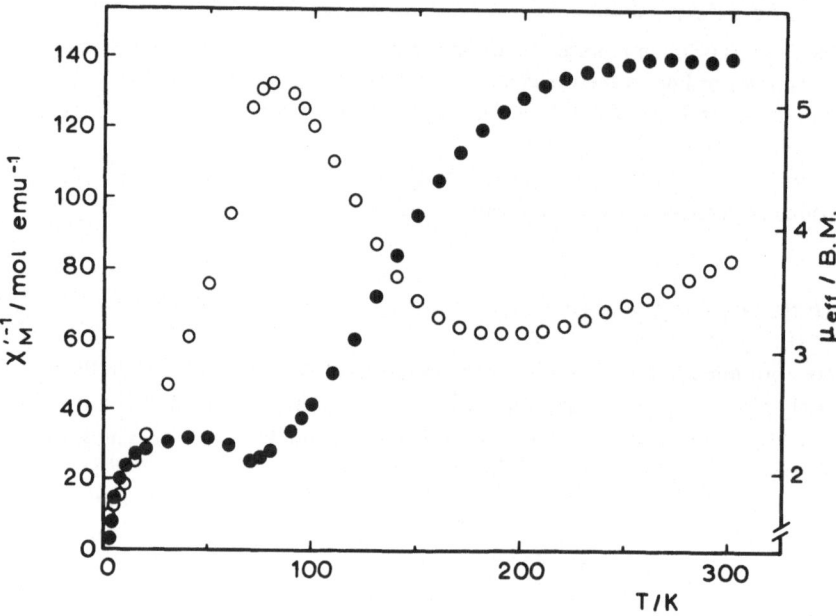

Fig. 14. Reciprocal molar magnetic susceptibility (○) and effective magnetic moment (●) of [Fe(2-CH$_3$-phen)$_3$](ClO$_4$)$_2$ as a function of temperature [102]

[Fe(2-CH$_3$-phen)$_3$]$^{2+}$ salts. The coexistence of the two spin states has been supported by Mössbauer measurements [102,103]. The spectra show well resolved quadrupole doublets due to the different spin states with quadrupolar precession times comparable with or longer than 10^{-7} s. In a thorough reinvestigation of the temperature dependence of the Mössbauer spectra of the perchlorate [102] a splitting of the spin quintet doublet into two doublets was observed for the first time, most likely arising from the ^{5}E and ^{5}A$_1$ states, which emerge from the ^{5}T$_2$(O$_h$) state by trigonal distortion (see Fig. 15). An Ingalls type [104] theoretical analysis of the temperature dependence of the quadrupole splittings $\Delta E_Q(^5E)$ and $\Delta E_Q(^5A_1)$, with trigonal field distortion, spin-orbit coupling, and covalency taken into account, gives results in accordance with this conclusion [102]. The amount of residual paramagnetism seems to be dependent on the sample preparation: From the area ratio under the Mössbauer resonance lines König et al. [103] found ~ 27% (at 4.2 K), Gütlich et al. [102] only 19% (at 4.2 K) of the molecules in the ^{5}T$_2$(O$_h$) state in case of the perchlorate. These figures vary with the nature of the anion [103].

The results from Mössbauer effect measurements on [Fe(2-CH$_3$-phen)$_3$] X$_2$ with X = I$^-$ and BPh$_4^-$ resemble very much those found for the perchlorate and tetrafluoroborate [106]. The application of an external magnetic field (40 kOe) yielded $V_{zz} > 0$ for the ^{1}A$_1$ state in all these salts [103,106].

At room temperature, the pure high-spin state may be partially (to about 40%) converted to the low-spin state on applying pressure of up to ~ 100 kbar, as has been demonstrated by Bargeron and Drickamer [98].

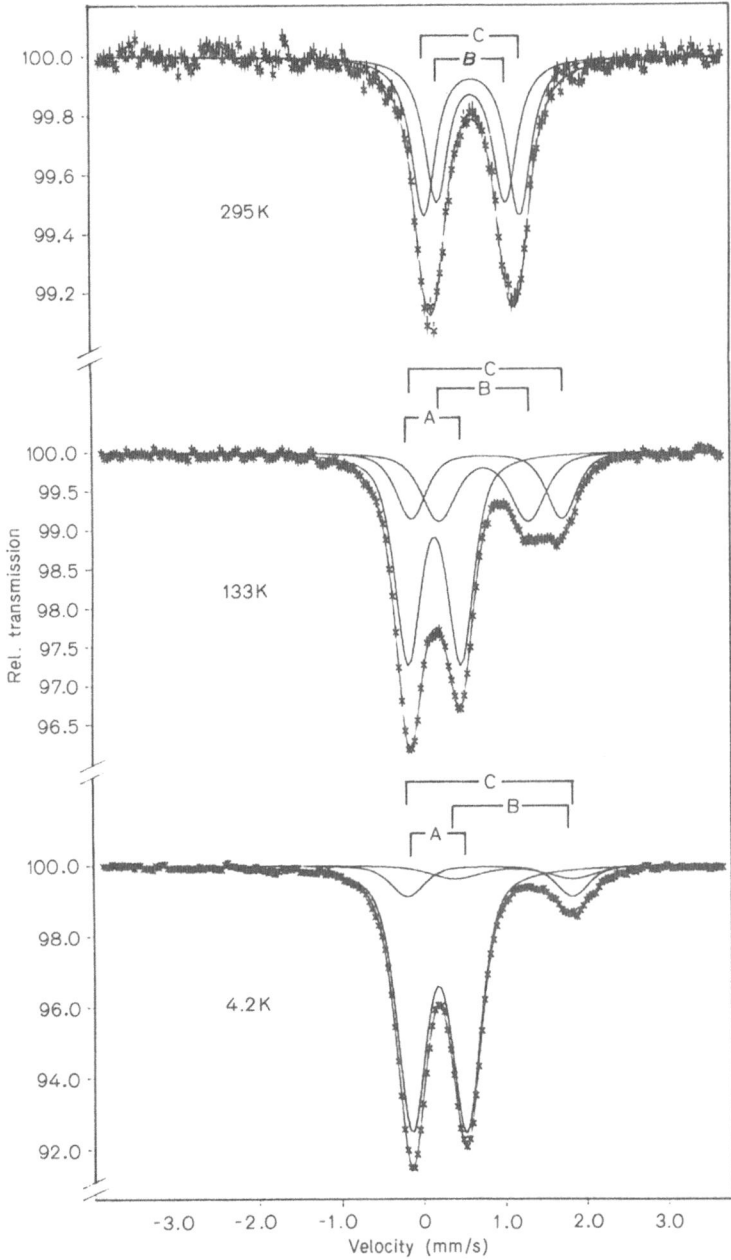

Fig. 15. Representative Mössbauer spectra of [Fe(2-CH$_3$-phen)$_3$](ClO$_4$)$_2$ at various temperatures; ^{57}Co/Cu source at 293 K. The three doublets are assigned to the ^{1}A$_1$ (A), ^{5}E (B), and ^{5}A$_1$ (C) states, serially, [102]

A frozen solution of $[Fe(2\text{-}CH_3\text{-}phen)_3](BF_4)_2$ in a methanol/glycerol mixture also exhibits high-spin $\rightleftharpoons$ low-spin transition between 89 and 160 K, as seen from variable temperature Mössbauer spectra [107]. The data of the isomer shift and the quadrupole splitting derived from the frozen solution spectra do not differ significantly from those of the crystalline material. The high-spin fraction, however, at comparable temperatures is some 20% higher in the frozen solution than in the pure solid. The authors thought this effect to be a specific property of the glass matrix. A concentration dependence of the spin transition behaviour in the frozen solution was not observed.

The compound $[Fe(2\text{-}CH_3O\text{-}phen)_3](ClO_4)_2 \cdot H_2O$, with a methoxy group in 2-position of phen also exhibits a gradual high-spin $(^5T_2)$ $\rightleftharpoons$ low spin $(^1A_1)$ transition, as has been established by magnetic studies and Mössbauer spectroscopy, down to 2.6 and 15 K, respectively [105]; cf. Figs. 16 and 17. Some representative magnetic data are collected in Table 4. Similar to the 2-CH_3-phen analog, a splitting of the high-spin quadrupole doublet into two doublets was observed, most likely due to the ^5E and $^5A_1$ crystal field states arising from the $^5T_2(O_h)$ state. The crossover behaviour deviates somewhat from that of the 2-CH_3-phen compound in that the high-spin state appears to be less favoured. Obviously, the ligand field potential seems to be slightly stronger in the 2-CH_3O-phen complex than in the 2-CH_3-phen complex. This may be a consequence of the fact that the 2-CH_3O-phen compound crystallizes with one H_2O molecule per unit formula, which could cause differences in the lattice packing and/or hydrogen bonding characteristics in comparison with the unhydrated

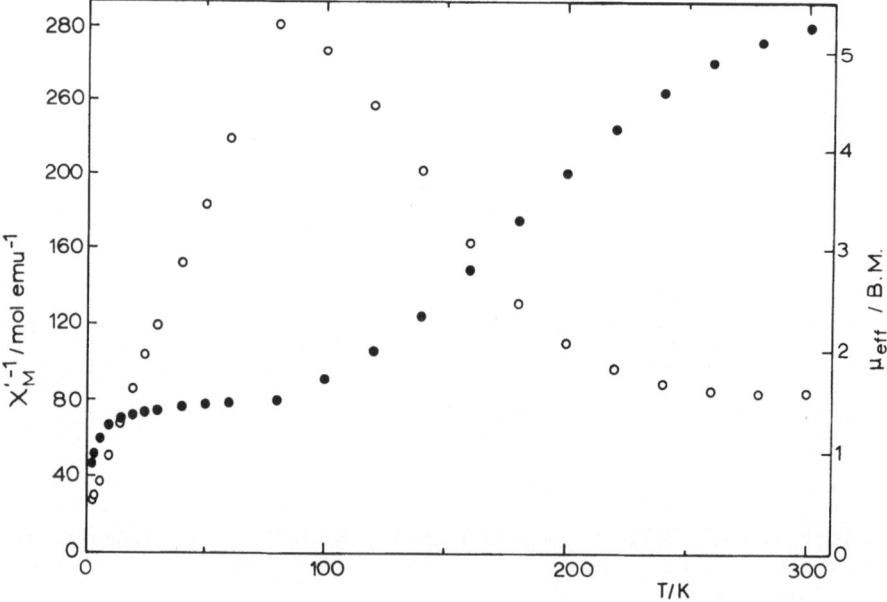

Fig. 16. Reciprocal molar magnetic susceptibility (o) and effective magnetic moment (•) of $[Fe(2\text{-}CH_3O\text{-}phen)_3](ClO_4)_2 \cdot H_2O$ as a function of temperature [105)]

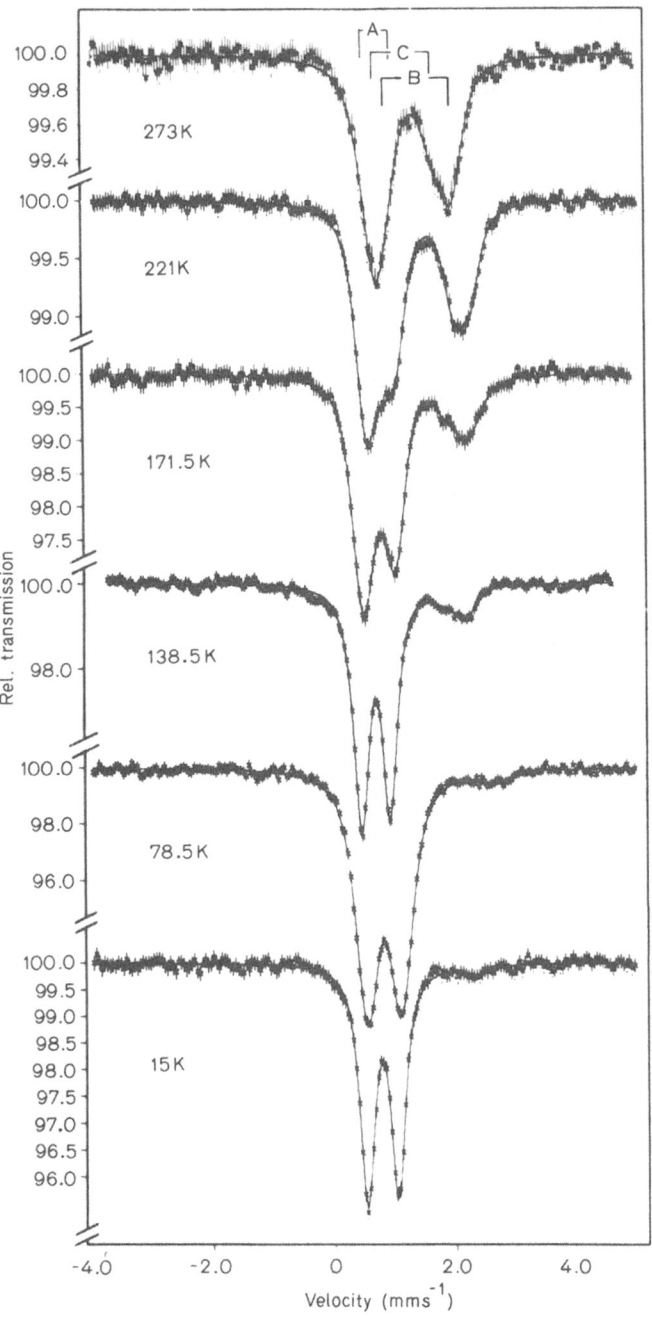

Fig. 17. Some representative Mössbauer spectra of $[Fe(2\text{-}CH_3O\text{-}phen)_3](ClO_4)_2 \cdot H_2O$ at various temperatures; the zero point of the velocity scale refers to the isomer shift of $Na_2[Fe(CN)_5NO] \cdot 2H_2O$ relative to the $^{57}Co/Cu$ source at 298 K. The doublets A, B and C are assigned to the $^1A_1$, $^5A_1$ and ^5E states, serially [105]

2-CH_3-phen compound. Such effects have indeed been observed in many spin crossover systems [109,110]. More likely, in our opinion, this behaviour seems to be the combined action of electronic and steric effects. The CH_3O substituent requires certainly more space than a methyl group, and one would therefore expect a more effective steric hindrance and thus a somewhat larger decrease of the partial ligand field strength in the 2-CH_3O-phen complex as compared to the 2-CH_3-phen complex. The observed overall effect, however, is a slight increase of 10 Dq in the 2-CH_3O-phen vs. the 2-CH_3-phen complex, which implies that most probably the electronic effect − here a somewhat stronger L → M σ-interaction − dominates over the steric effect. Results from some empirical CNDO/2 calculations [105] support this interpretation.

The [Fe(2-Cl-phen)$_3$]$^{2+}$ complex ion was originally also thought to exhibit spin crossover as concluded from a slight decrease of the magnetic moment between room temperature and 102 K [111]. Later, variable-temperature Mössbauer measurements, independently carried out by Reiff and Long [69] and Gütlich et al. [112,113], provided clear evidence that the 2-Cl-phen complex is purely high-spin down to 4.2 K. The Mössbauer spectra show a single quadrupole doublet at all temperatures, with typical iron (II) high-spin parameter values. Mössbauer measurements under applied magnetic field [69] have established the ^5A term as the ground state. The same result was arrived at by a theoretical treatment of the temperature dependence of the quadrupole splitting [112]. Undoubtedly, it is the relatively high electronegativity of the chlorine atom, which reduces the basicity of the coordinating nitrogen atom, with the expected consequence that both L → M σ-donation and M → L π-backbonding will be reduced and thus the ligand field strength decreased. The decrease in the field strength apparently is so large, that 10 Dq is shifted significantly away from the crossover point (Δ_c). The steric effect, still being operative in comparison with the unsubstituted phen ligand, obviously plays a minor role.

The following order of substituents Y in 2-Y-phen ligands with decreasing ligand field strengths (and thus increasing tendency towards paramagnetic behaviour) of the [Fe(2-Y-phen)$_3$] X$_2$ complexes can be constructed: Y=H > CH_3O > CH_3 > Cl. There is no systematic tendency involving the size of these substituents in the given ordering. This implies that the size effect alone does not govern the magnetic behaviour of the corresponding iron(II) complexes. It appears that only in the step from Y=H to any of the other substituents does the steric effect play a decisive role; within the non-hydrogen substituents, however, the electronic effect, i.e. gradation in the sum of (M−L) σ- and π-interaction, seems to dominate. This has been supported by results from semiempirical CNDO/2 calculations [105,114], which, although expressing only tendencies in the changes of molecular orbital energies, show how the basicity of the coordinating nitrogen atom varies with the nature of the substituent Y in 2-Y-phen. The CNDO/2 results are fully consistent with the above ordering of Y substituents.

The [Fe(2,9-$(CH_3)_2$-phen)$_3$] $(ClO_4)_2$ is paramagnetic, most likely due to the enhanced steric effect on going from the unsubstituted phen to the 2,9-$(CH_3)_2$-phen ligand [114].

4.2 Other Tris(substituted-phenanthroline) Complexes

Epstein [115] measured the ^{57}Fe Mössbauer spectra of a number of tris-complexes of iron (II) (as perchlorates) with substituted phenanthroline ligands such as 5-NO_2-phen, 5-Cl-phen, 5-CH_3-phen, 5,6-$(CH_3)_2$-phen, 4,7-$(CH_3)_2$-phen, 3,4,7,8-$(CH_3)_4$-phen, 5-C_6H_5-phen, and found that all these complexes behave very similarly to the unsubstituted parent system [Fe(phen)$_3$](ClO$_4$)$_2$ with a $^1A_1$ ground state. Both the isomer shift and the quadrupole splitting in these low-spin complexes do not differ significantly from the corresponding data of [Fe(phen)$_3$] (ClO$_4$)$_2$. Clearly, both the electronic effects and the steric hindrance, particularly in the remote positions of phen (with respect to the iron center) are not strong enough to decrease the ligand field strength sufficiently to make $|\Delta-\pi|$ comparable to kT and eventually exhibit spin crossover. A few of these complexes, and in addition some other tris(substituted-phen) complexes, have been studied using Mössbauer spectroscopy by Collins, Pettit, and Baker [116], with very similar results. No indication of a spin change was observed. Differences in the quadrupole splitting could be correlated with the nephelauxetic effect of the ligands.

The class of [Fe(4-Y-phen)$_3$] (ClO$_4$)$_2 \cdot$ nH$_2$O systems with Y = CH_3, Cl, C_2H_5COO, CHNOH, CN, and the class of [Fe(4,7-Y$_2$-phen)$_3$] (ClO$_4$)$_2 \cdot$ nH$_2$O systems with Y = CH_3, Cl, C_2H_5COO, all show low-spin behaviour in a Mössbauer effect study [117], with δ- and ΔE_Q values typical of iron (II) in the $^1A_1$ ground state.

Thus the only spin crossover systems of the type [Fe(substituted-phen)$_3$] X$_2 \cdot$ nH$_2$O known at the present time (and to the best of our knowledge) are the 2-CH_3-phen series with X = I$^-$, BF$_4^-$, PF$_6^-$, and ClO$_4^-$, and the [Fe(2-CH_3O-phen)$_3$] (ClO$_4$)$_2 \cdot$ H$_2$O complex.

5 Complexes with Tridentate (2-substituted-phenanthroline) Ligands

Goodwin and his school have investigated the effect on magnetism of the incorporation of a donor-atom containing substituent into the 2-position of phenanthroline producing a tridentate chelating ligand phen-2-Y. Depending on the kind of the co-ordinating atom in the substituent Y, one may classify the hexa-coordinate iron(II) complexes with respect to the resulting core Fe(NND)$_2$ where D is the donor-atom of the substituent Y.

phen-2-Y

5.1 Complexes with Fe(NNO)$_2$ Core

Simple derivatives of this kind are 2-carboxy-1,10-phenanthroline (phen-2-COO$^\ominus$) (I) and 1,10-phenanthroline-2-carboxamide (phen-2-CONH$_2$) (II).

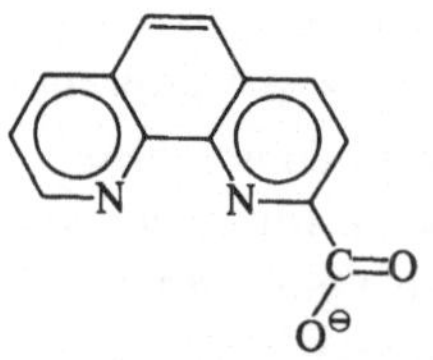

(I) 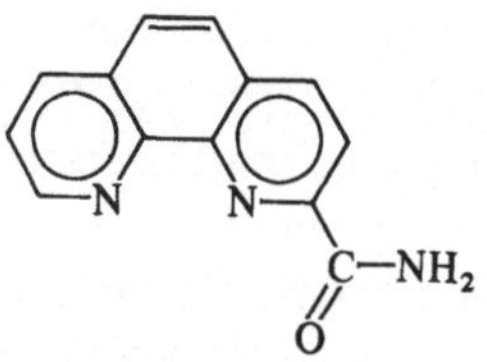   (II)

The iron(II) chelates [Fe(phen-2-COO)$_2$] · 1,5 H$_2$O [118] and [Fe(phen-2-CONH)$_2$] X$_2$ · nH$_2$O with X = ClO$_4^-$ (n = 2), BF$_4^-$ (n = 2), NO$_3^-$ (n = 2), and Br$^-$ (n = 3) [111,119] are paramagnetic and follow well the Curie-Weiss law. Some magnetic data are collected in Table 5. The drastic change from the low-spin behaviour of the Fe(NN)$_3$ core of [Fe(phen)$_3$]$^{2+}$ to the high-spin behaviour of the Fe(NNO)$_2$ core of the present bis(tridentate) chelates is undoubtedly due to the poor π-bonding properties of the oxygen in comparison with those of the nitrogen donor atom. A tris-ligand iron(II) complex of phen-2-CONH$_2$ has also been described [119]. This, too, is high-spin, although the bidentate amide most likely produces a somewhat stronger ligand field than the tridentate ligand.

5.2 Complexes with Fe(NNS)$_2$ Core

Complexes of this kind are [Fe(phen-2-CSNH$_2$)$_2$] X$_2$ · nH$_2$O, in which the tridentate ligand, 1,10-phenanthroline-2-carbothioamide (phen-2-CSNH$_2$) (III) is expected to

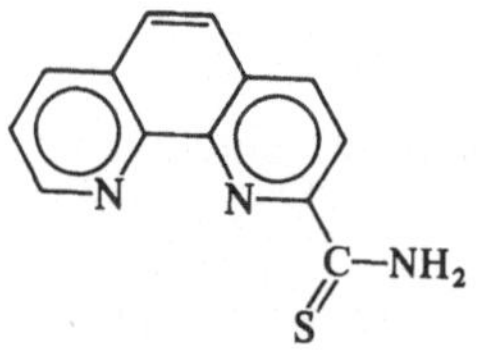

(III)

produce a considerably stronger ligand field than the analogous phen-2-carboxamide ligand, because of the better π-bonding capability of the sulfur donor atoms in the Fe(NNS)$_2$ core. The complexes with various anions X are all low-spin with magnetic moments around 0.7–0.8 B.M. at room temperature (see Table 5). The neutral complex [Fe(phen-2-CSNH)$_2$] is also diamagnetic [120]. The same is true for the N-phenyl substituted complex [Fe(phen-2-CSNH(Ph))$_2$] [121]. Obviously, the critical field strength Δ_c (crossover point) lies in between the oxygen and the sulfur donor atom in corresponding complexes with Y$_O$ = CONH$_2$ and Y$_S$ = CSNH$_2$, respectively. The distances from the crossover point, expressed as $|\Delta(Y_O)-\Delta_c|$ and $|\Delta(Y_S)-\Delta_c|$ apparently are well above kT so that the energetically higher spin state, HS and LS, respectively, is not thermally populated to any measurable extent.

Table 5. Magnetic data of $[Fe(phen-2-Y)_2] \cdot nH_2O$ and $[Fe(phen-2-Y)_2]X_2 \cdot nH_2O$ complexes with tridentate phenanthroline derivatives

Compound	T(K)	μ_{eff} (B.M.)	Ground state[a]	Ref.
$[Fe(phen-2-COO)_2] \cdot 1.5H_2O$	298/205/118	5.28/5.25/ 5.26	$^5T_2$	118
$[Fe(phen-2-CONH_2)_2]X_2 \cdot nH_2O$				
X = ClO_4^- (n = 2)	303/83	5.3/4.9	$^5T_2$	119
X = BF_4^- (n = 2)	303/83	5.3/5.1	$^5T_2$	119
X = NO_3^- (n = 2)	303/83	5.4/5.2	$^5T_2$	119
X = Br^- (n = 3)	303/83	5.4/5.2	$^5T_2$	119
$[Fe(phen-2-CSNH_2)_2]X_2 \cdot nH_2O$				
X = Cl^- (n = 1)	295	0.7	$^1A_1$	120
X = I^- (n = 2)	295	0.8	$^1A_1$	120
X = NO_3^- (n = 0)	295	0.8	$^1A_1$	120
X = ClO_4^- (n = 1)	295	0.7	$^1A_1$	120
$X_2 = SO_4^{2-}$ (n = 1)	295	0.8	$^1A_1$	120
$[Fe(phen-2-CSNH_2)_2]$	295	0.9	$^1A_1$	120
$[Fe(phen-2-C(=NOH)NH_2)_2]-$ $ClO_4 \cdot H_2O$	298	0	$^1A_1$	122
$[Fe(phen-2-CH=N-NR^1R^2)_2]-$ $X_2 \cdot nH_2O$				
$R^1 = H, R^2 = Ph, X = BF_4^-$ (n = 0)	343/291/287/ 279/263/83	5.28/5.16/ 4.66/1.62/ 1.18/0.61	$^5T_2 \rightleftharpoons {}^1A_1$	123
$R^1 = H, R^2 = CH_3, X = BF_4^-$ (n = 0.5)	363/283/83	1.92/1.66/ 1.42	$^1A_1 (\rightleftharpoons {}^5T_2)$	123
$R^1 = CH_3, R^2 = CH_3, X = BF_4^-$ (n = 1)	313/193/83	5.51/5.15/ 4.41	$^5T_2 (\rightleftharpoons {}^1A_1)$	123
$R^1 = CH_3, R^2 = C_6H_5, X = BF_4^-$ (n = 1)	303/83	5.1/5.1	$^5T_2$	123
$R^1 = C_6H_5, R^2 = C_6H_5, X = BF_4^-$ (n = 0.5)	303/83	5.0/5.0	$^5T_2$	123
$R^1 = H, R^2 = 2\text{-pyridyl}, X = BF_4^-$ (n = 1)	303/83	5.2/4.8	$^5T_2$	123
$[Fe(phen-2-CH=NR)_2](BF_4)_2 \cdot nH_2O$				
$R = Bu^t$, n = 1	343/243/183/ 123/83	5.12/4.18/ 3.14/2.31/ 1.96	$^5T_2 \rightleftharpoons {}^1A_1$	125
$R = Me, Et, Pr, Pr^i, Bu, Bu^s;$ Ph, Benzyl	–	–	$^1A_1$	125
$[Fe(phen-2-VII)_2]X_2;$				
X = BF_4^-	303[b]/303[c] 83[b]/83[c]	5.34/5.13 3.63/1.89	$^5T_2 \rightleftharpoons {}^1A_1$	126
X = ClO_4^-	303[b]/303[d] 83[b]/83[d]	5.32/5.38 3.61/3.55	$^5T_2 \rightleftharpoons {}^1A_1$	126
X = I^-	303[b]/303[d] 83[b]/83[d]	5.50/4.91 3.32/1.89	$^5T_2 \rightleftharpoons {}^1A_1$	126
$[Fe(phen-2-VIII)_2](BF_4)_2 \cdot H_2O$	363/333/303/ 273/203/83	3.26/2.57/ 1.81/1.39/ 0.99/0.76	$^5T_2 \rightleftharpoons {}^1A_1$	128
$[Fe(phen-2-IX)_2](BF_4)_2 \cdot 2H_2O$	373/273/93	1.72/1.10/ 0.81	$^1A_1 (\rightleftharpoons {}^5T_2)$	128

Table 5. (continued)

Compound	T(K)	μ_{eff} (B.M.)	Ground state[a]	Ref.
[Fe(phen-2−IX−H)$_2$]·0.5 CHCl$_3$	363/303/83	1.79/0.96/ 0.70	$^1A_1$ ($\rightleftharpoons$ $^5T_2$)	128
[Fe(phen-2−X)$_2$](BF$_4$)$_2$·2H$_2$O	343/303/203/ 113/83	3.66/3.01/ 2.05/1.66/ 1.60	$^5T_2$ $\rightleftharpoons$ $^1A_1$	127
[Fe(phen-2−XI)$_2$]X$_2$·nH$_2$O				
X = I$^-$ (n = 3)	303/83	5.2/5.0	$^5T_2$	127
X = Br$^-$ (n = 4)	303/83	5.4/5.3	$^5T_2$	127
X = BF$_4^-$ (n = 3)	303/83	5.2/5.1	$^5T_2$	127
X = NO$_3^-$ (n = 4)	303/83	5.4/5.4	$^5T_2$	127
[Fe(phen-2−XII)$_2$](BF$_4$)$_2$·H$_2$O	363/243/83	4.8/4.62/ 4.20	$^5T_2$ ($\rightleftharpoons$ $^1A_1$)	127
[Fe(phen-2−XIII)$_2$](BF$_4$)$_2$·H$_2$O	363/303/203/ 103/83	3.92/2.81/ 1.89/1.68/ 1.58	$^5T_2$ $\rightleftharpoons$ $^1A_1$	127

[a] Term symbols given in the approximation of O$_h$ symmetry
[b] Measured immediately after preparation
[c] After 50 weeks
[d] After 6 weeks

Goodwin and his coworkers have played the game further with the incorporation of donor-atom containing substituents in the 2-position of phen. Only with some nitrogen-donor containing substituents were they able to move the ligand field strength at iron (II) sufficiently close to the critical value Δ_c to create spin crossover, as will be described next.

5.3 Complexes with Fe(NNN')$_2$ Core

The ligand 1,10-phenanthroline-2-amidoxime (phen-2-C(=NOH)NH$_2$) (IV) still produces a fairly strong field strength, and the iron (II) complex [Fe(phen-2-C(=NOH)-NH$_2$)]$_2$ (ClO$_4$)$_2$·H$_2$O is diamagnetic (μ_{eff} = 0 B.M. at 298 K), whereas the corresponding nickel(II) and cobalt(II) complexes have high-spin ground states [122].

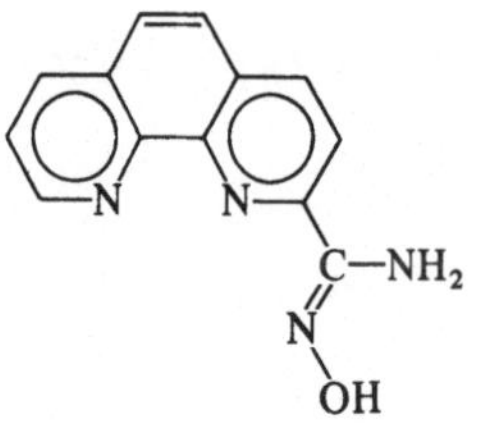

(IV) 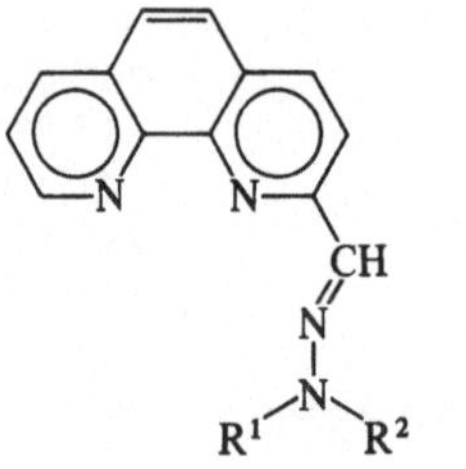   (V)

The ligand field strength approaches the spin crossover point in the iron(II) complexes of 1,10-phenanthroline-2-carbaldehyde hydrazones, (phen-2-CH=N-NR^{1}R^2) (V) [123].

The ligand (V) shows a gradation in field strength with the varying nature of R^1 and R^2; this is reflected in the spin state of the iron complexes [Fe(phen-2-CH=N-NR^{1}R^2)$_2$] (BF$_4$)$_2 \cdot$nH$_2$O (see Table 5). The most pronounced change in magnetism occurs in the complex with R^1 = H, R^2 = Ph. The magnetic moment changes very abruptly near room temperature, probably indicating a first-order transition similar to that in [Fe(phen)$_2$(NCS)$_2$]. The values of μ_{eff} at both the high-temperature and the low-temperature limit are very close to what one expects generally for pure S = 2 and S = 0 states, respectively. So the transition is virtually complete. The complex with R^1 = H, R^2 = CH$_3$ most likely is low-spin ($^1A_1$); however, a spin crossover occuring above room temperature must be considered possible on the basis of the present magnetic data. At any rate, the field strength is lower in case of a methyl group for R^2 instead of a phenyl group. Again, differences in steric hindrance and in electronic effects may be responsible for the different spin state behaviour. The steric effect dominates in the high-spin complexes with a bulky phenyl group together with a methyl group or with two phenyl groups for R^1 and R^2.

To see whether structural changes eventually accompany the temperature dependent changes in the spin state in [Fe(phen-2-CH=N-N(H) (C$_6$H$_5$)$_2$] (ClO$_4$)$_2$, X-ray powder diffraction patterns have been measured, particularly in the neighbourhood of T$_c$ = 263 K [124]. Two different diffraction patterns are obtained above and below T$_c$, which demonstrates that the $^5T_2 \rightleftharpoons {}^1A_1$ transition is accompanied by a crystallographic change. The peak profiles of the X-ray powder diffraction show the same temperature dependence as the $^5T_2$ and $^1A_1$ fractions determined from Mössbauer and magnetic measurements. This indicates that the crystallographic change is not due to a continuous molecular distortion, but is directly associated with the change of the spin state. The peak profiles have also been found to follow accurately the hysteresis behaviour which most likely reflects the existence of domains of like spin rather than a statistical distribution of molecules with $^5T_2$ and $^1A_1$ ground states, respectively.

The magnetic moments of the tetrafluoroborate complexes with R^1 = H, R^2 = CH$_3$ and R^1 = CH$_3$, R^2 = CH$_3$ in nitromethane solution at 290 K were found to be 1.1 and 5.6 B.M., respectively, in close agreement with the values for the solid samples at the same temperatures [123]. These results demonstrate that uncontrollable lattice effects do not play a crucial role here, and that the actual field strengths are mainly determined by the ligand properties alone.

Similar effects of intra-ligand substitution on the ligand field strength have been observed on iron (II) complexes with 1,10-phenanthroline-2-carbaldehyde imines (phen-2-CH=NR) (VI) [125]. The only complex which has been found to exhibit spin crossover (gradual in this case) is the tetrafluoroborate with R = tertiary butyl (Bu^t)

(VI)

[Fe(phen-2-CH=NBu^t)$_2$] (BF$_4$)$_2$ · H$_2$O; cf. Table 5. All other complexes with R = alkyl (methyl, ethyl, propyl, isopropyl, butyl, s-butyl) or R = aryl (phenyl, benzyl) are spin-paired at room temperature with ^{1}A$_1$ ground states [125]. From the electronic spectra of the nickel complexes an ordering of field strengths has been established which correlates with the bulkiness of the imine moiety. It is interesting to note that certain of the ligands of high field strength (R = methyl, ethyl, propyl, isopropyl) exert a gradual ^{4}T$_1$(O$_h$) ⇌ ^{2}E(O$_h$) transition in cobalt(II) complexes.

Goodwin gives an instructive comparison of the magnetic properties of the iron (II) complexes with closely related phen-2-CH=NR ligands, which shows how sensitive these are to minor structural changes, when the field strength is in the crossover region [14]; R = C$_6$H$_5$CH$_2$ induces low-spin behaviour, R = C$_6$H$_5$NH gives rise to spin crossover, and R = (C$_5$H$_5$N)NH causes high-spin behaviour.

Other ter-imine complexes with magnetically interesting behaviour, prepared and studied by the Goodwin school, contain a five-membered heterocycle with a nitrogen-donor atom [126,127,128] in the 2-position of 1,10-phenanthroline.

The red complexes [Fe(phen-2-VII)$_2$] X$_2$, with X = BF$_4^-$, ClO$_4^-$, I$^-$ and phen-2-VII is 3-(2-(1,10-phenanthrolyl))-5-methyl-1,2,4-oxadiazole, are reported to be high-spin at ambient temperature, but undergo a gradual spin transition on lowering

(phen-2-VII)

the temperature [126]. The transition is not complete and a large residual paramagnetism is observed. The main feature of this system is the time-dependence of the fraction of the residual paramagnetism over a period of about 50 weeks. We are possibly dealing with a "crystal aging" effect here. X-ray powder photographs did not show any structural change in the crystal lattice [126].

The complex [Fe(phen-2-VII)$_2$] X$_2$, with X = BF$_4^-$, ClO$_4^-$, I$^-$ and phen-phenanthroline-2-yl) imidazoline, is predominantly low-spin at room temperature with a magnetic moment of ~ 1.8 B.M. [128]. Above room temperature, however, μ_{eff} rises gradually, and a smooth ^{1}A$_1$(O$_h$) ⇌ ^{5}T$_2$(O$_h$) transition is to be considered in this case.

(phen-2-VIII)

The benzimidazole analog, [Fe(phen-2-IX)$_2$] (BF$_4$)$_2$ · 2H$_2$O, and its neutral complex [Fe(phen-2-IX-H)$_2$] · 0.5 CHCl$_3$ are both low-spin at room temperature [128].

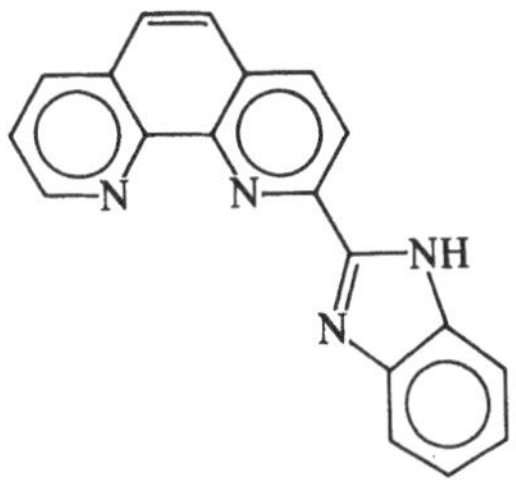

(phen-2-IX)

The magnetic moments rise only slightly at elevated temperatures (see Table 5), which led the authors to conclude that some population of the higher $^5T_2(O_h)$ state is possible. No clear distinction can be made as to which of the influencing factors, viz. electronic effects, steric hindrance, and crystal solvent effects, plays the dominant role here, because all of these are operative to some extent. Data from the UV-vis spectra of the nickel(II) complexes indicate that the ligands have field strengths in the iron(II) crossover region.

In the crossover region are also iron(II) complexes with 2-(1,10-phenanthroline-2-yl)thiazole derivatives. Goodwin et al. [127] have suggested that in $[Fe(phen-2-X)_2]$ $(BF_4)_2 \cdot 2H_2O$ (phen-2-X is 2-(1,10-phenanthroline-2-yl)thiazole), in $[Fe(phen-2-XII)_2]$ $(BF_4)_2 \cdot H_2O$ (phen-2-XII is 2-(1,10-phenanthroline-2-yl)thiazolidine), and in $[Fe(phen-2-XIII)_2](BF_4)_2 \cdot H_2O$ (phen-2-XIII is 2-(1,10-phenanthroline-2-yl)benzo-thiazole) a smooth temperature-dependent $^5T_2(O_h) \rightleftharpoons {}^1A_1(O_h)$ transition takes place, which in none of these complexes is complete within the experimental temperature range (83—363 K). The pyridylthiazole analogs, $[Fe(phen-2-XI)_2]X_2 \cdot nH_2O$ (see Table 5) are high-spin throughout the range 83—363 K, which the authors ascribe to the uncoordinated pyridyl group hindering the close approach of ligand and metal atoms necessary for spin-pairing.

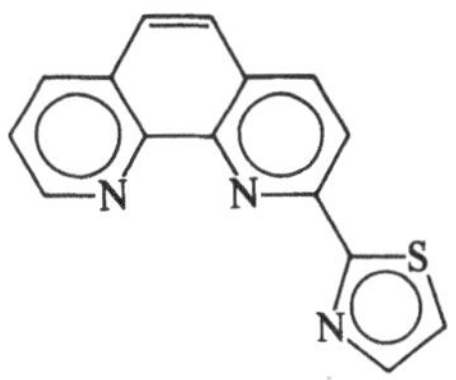

(phen-2-X)

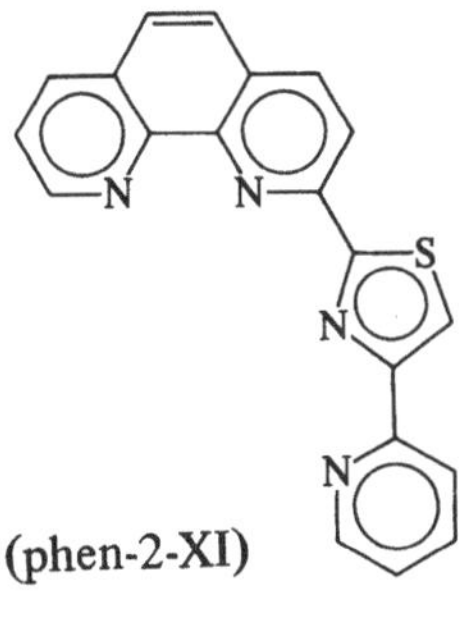

(phen-2-XI)

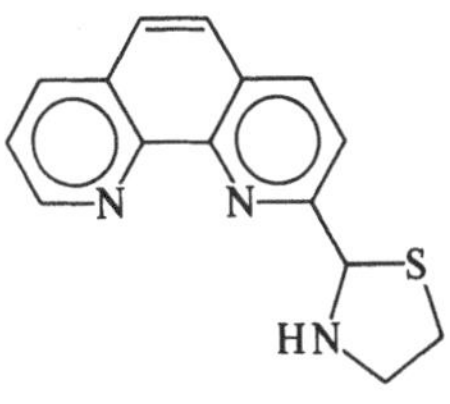

(phen-2-XII)

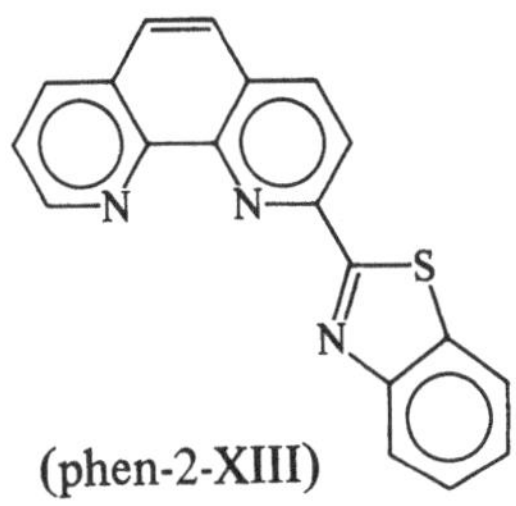

(phen-2-XIII)

6 Complexes with Bidentate (2-substituted-pyridine) Ligands

6.1 Tris(2-picolylamine) Complexes

Sutton [129] communicated in 1960 that the $[Fe(2\text{-}pic)_3] X_2$ salts are paramagnetic at room temperature with magnetic moments expected for $^5T_2(O_h)$ ground states.

Seven years later, Renovitch and Baker [130] measured the magnetic susceptibility of the chloride, bromide, and iodide over the temperature range 20–300 K and found that the effective magnetic moment depends strongly on the nature of the uncoordinated anion (see Fig. 18). The $\mu_{eff}(T)$ curves provide clear evidence for a $^5T_2(O_h) \rightleftharpoons {}^1A_1(O_h)$ transition in the chloride and the bromide, although they differ significantly in the extent of residual paramagnetism in the low-temperature

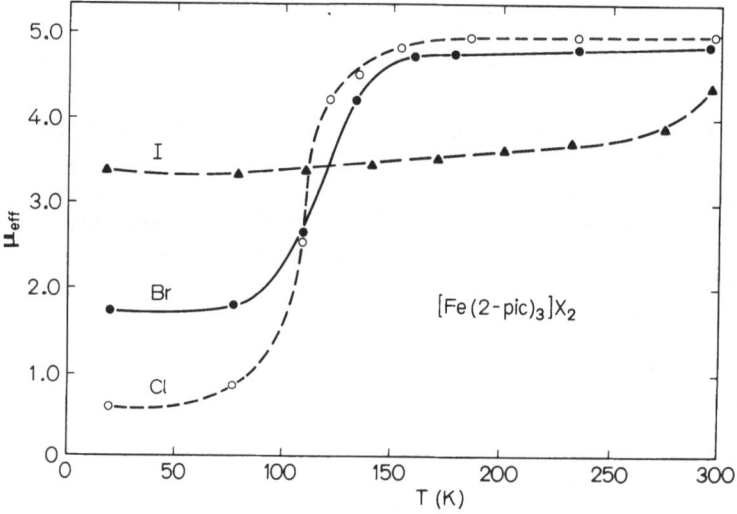

Fig. 18. Temperature dependence of the effective magnetic moment μ_{eff} of $[Fe(2\text{-}pic)_3]X_2$, $X = Cl^-, Br^-, I^-$ (from Ref. 130)

region. The situation with the iodide appears not to be as clear from the $\mu_{eff}(T)$ measurements; one may rather be tempted to interpret its magnetism with a dominating intermediate-spin ground state. However, the Mössbauer spectra of the chloride and the iodide (the bromide gave only poor resonance effects due to self-absorption of the 14.4 keV γ-radiation) revealed the presence of two doublets in each case, with parameters typical of $^5T_2(O_h)$ and $^1A_1(O_h)$ ground states. The temperature dependence of their intensities agree with the magnetic data. The authors also performed magnetic and Mössbauer measurements on the zinc diluted solid solution $[Fe_{0.15}Zn_{0.85}(2\text{-}pic)_3] Cl_2$; the results were identical with those of the pure iron system, which led the authors to the conclusion that antiferromagnetic interaction is not the cause of the anomalous magnetic behaviour.

6.1.1 Effect of Metal Dilution

A systematic study of the metal dilution effect on the spin-crossover behaviour in the solid solutions $[Fe_xZn_{1-x}(2\text{-pic})_3]Cl_2 \cdot EtOH$ $(0.0009 \leqslant x \leqslant 1)$ has been performed employing ^{57}Fe Mössbauer spectroscopy between 5 and 300 K [84,85]. The purpose was to find support for the cooperative domain model suggested earlier by Sorai and Seki [34,87]. Some representative Mössbauer spectra of the undiluted system $(x = 1)$ as a function of temperature are displayed in Fig. 19. The spectra demonstrate that

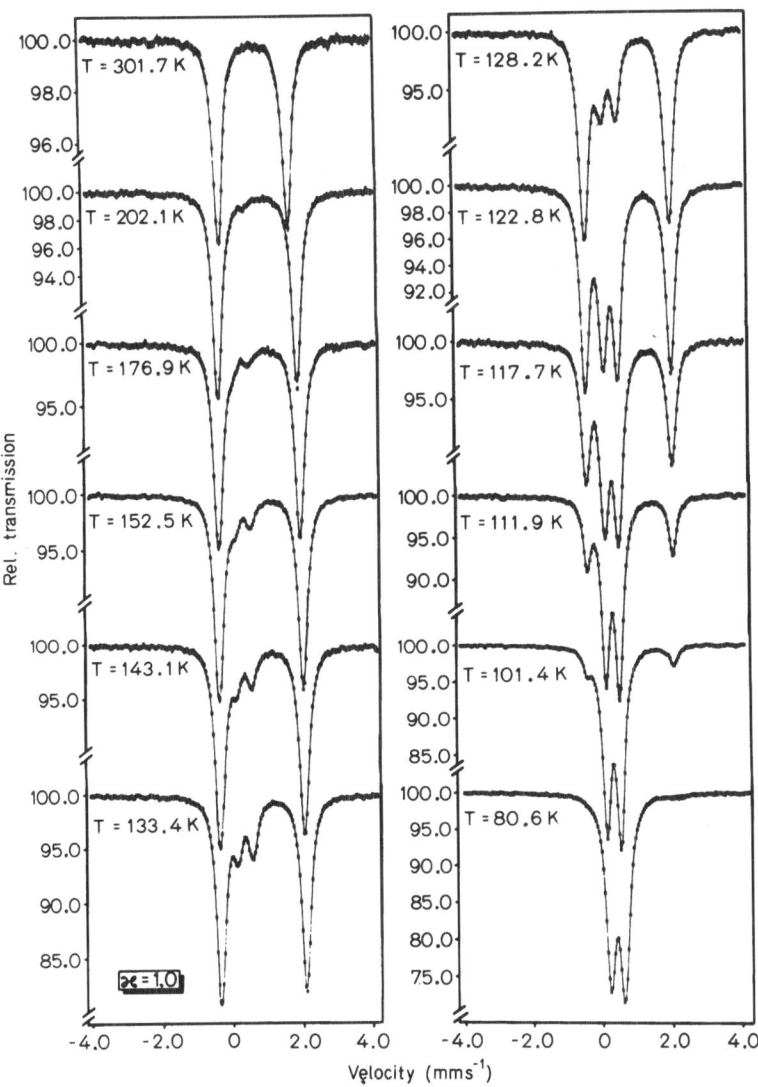

Fig. 19. ^{57}Fe Mössbauer spectra of $[Fe(2\text{-pic})_3]Cl_2 \cdot EtOH$ at various temperatures [84]

the transition is complete in both the high-temperature and the low-temperature regions, viz. the spectra near 300 K consist of a single quadrupole doublet arising from the $^5T_2(O_h)$ state, and around 80 K the spectra show a single quadrupole doublet due to the $^1A_1(O_h)$ state. The concentration dependence of the relative intensities is illustrated in Fig. 20, which clearly states that, with decreasing iron

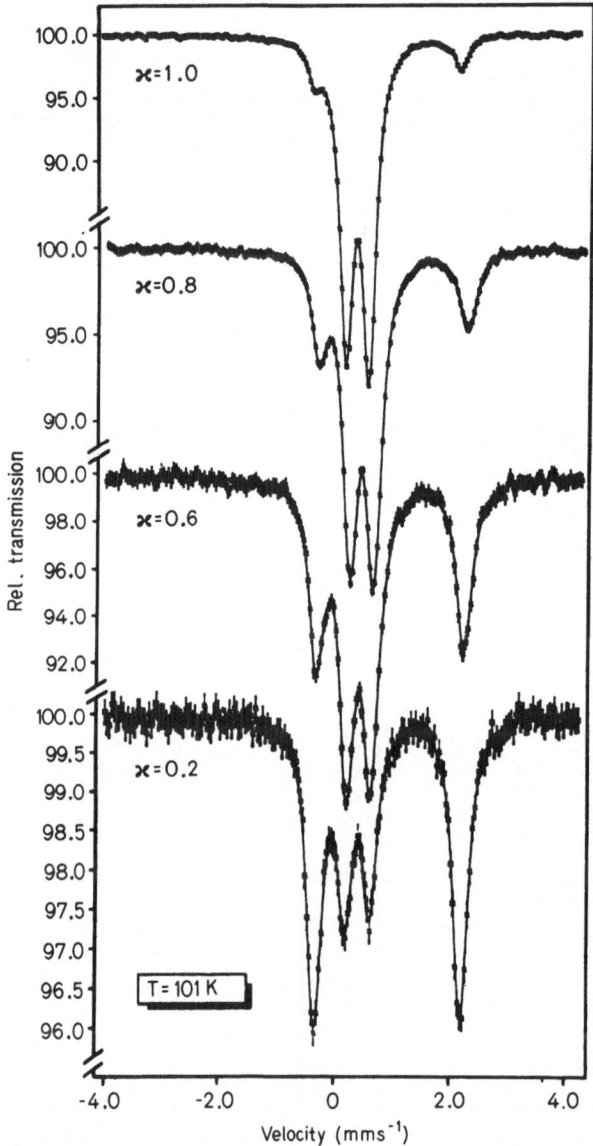

Fig. 20. Concentration dependence of ^{57}Fe Mössbauer spectra at 101 K for [Fe$_x$Zn$_{1-x}$(2-pic)$_3$]Cl$_2 \cdot$ EtOH [84)]

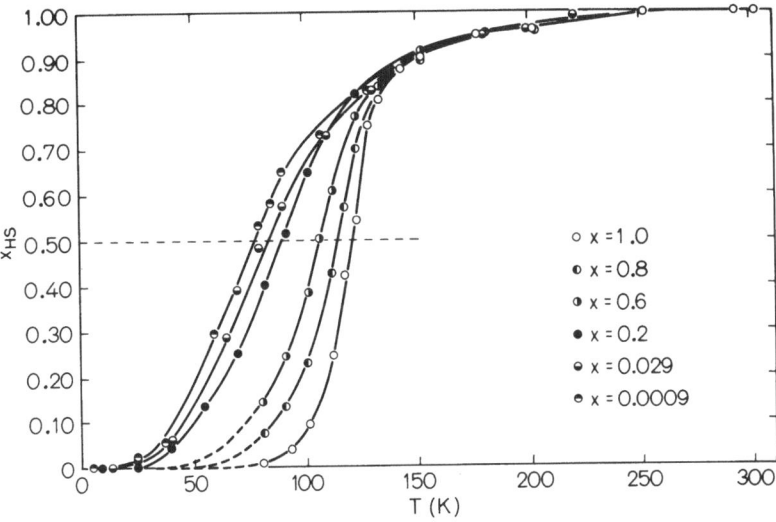

Fig. 21. Temperature dependence of the area fraction x_{HS} of the high-spin quadrupole doublet in $[Fe_xZn_{1-x}(2\text{-pic})_3]Cl_2 \cdot EtOH$ for variable iron concentration x [84]

concentration in the zinc host, the $^5T_2(O_h)$ becomes more and more stabilized over the $^1A_1(O_h)$ state. This is further visualized in Fig. 21, where the area fraction of the high-spin doublet is plotted as a function of temperature for different iron concentrations x. The $x_{HS}(T)$ curves become more gradual and are shifted towards lower temperatures with increasing dilution of iron in the zinc host. The transition temperature $T_c(x)$ (taken as the temperature at $x_{HS} = 0.5$) varies nearly linearly with the iron concentration x (see Fig. 22) in the range $1.0 \geqslant x \geqslant 0.15$.

This characteristic $T_c(x)$ behaviour has been considered a massive support for the cooperative domain model of Sorai and Seki [34,87] suggesting that the spin transition takes place through a coupling between the electronic state and vibrational modes, and that the conversion of the electronic state occurs simultaneously in a group of molecules, which form a so-called "cooperative domain (region)". The results obtained from the metal dilution studies on $[Fe_xZn_{1-x}(2\text{-pic})_3] Cl_2 \cdot EtOH$, and more recently on $[Fe_xCo_{1-x}(2\text{-pic})_3] Cl_2 \cdot EtOH$ [108] and $[Fe_xMn_{1-x}(phen)_2(NCS)_2]$ [179], lend support to the conclusion that the cooperative domain model may eventually be generalized to spin-crossover systems of both types, smooth (gradual) and abrupt. A major difference would be the size of the cooperative domain, e.g. comprising on the order of 100 complex molecules in a "strongly coupling" system with abrupt spin change such as in $[Fe(phen)_2(NCS)_2]$ [34], or much less molecules in "weakly coupling" systems such as the metal-diluted ones discussed here, with low iron concentrations, or pure iron spin-crossover systems with gradual spin conversion [86]. On the basis of these ideas a phenomenological thermodynamic model has been applied successfully to interpret the spin transition in polycrystalline solid solutions [86]. The model allows one to calculate the spin transition temperature T_c using the effective

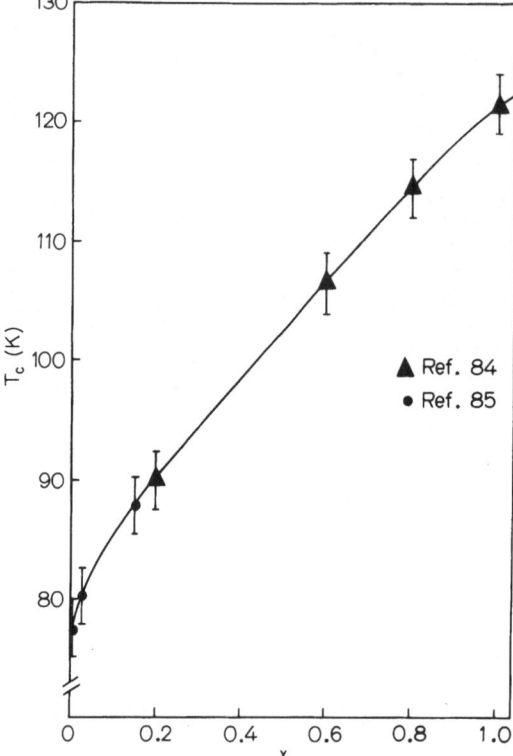

Fig. 22. Concentration dependence of the spin transition temperature T_c for $[Fe_xZn_{1-x}(2\text{-pic})_3]Cl_2 \cdot EtOH$ [84,85]

changes of enthalpy and entropy, respectively, obtained from lnK vs. 1/T plots (where $K = x_{HS}/1-x_{HS}$). Phenomenological expressions for the transition temperature $T_c(x)$ as well as for the effective enthalpy and entropy changes $\Delta H_{eff}(x)$ and $\Delta S_{eff}(x)$, on going from low-spin to high-spin, as a function of concentration x have been adjusted to the measured data. An average domain size of about 4 complex molecules was found for the pure iron complex $[Fe(2\text{-pic})_3]Cl_2 \cdot EtOH$, which indicates comparatively weak intermolecular coupling. The domain size depends markedly on the iron concentration in the solid solutions $[Fe_xZn_{1-x}(2\text{-pic})_3]Cl_2 \cdot EtOH$, and approaches n = 1 at very low concentration x. In the pure iron compound, the sum of the intramolecular enthalpy contributions (mainly electronic and vibrational) changes by some 2300 J mol^{-1} and that of the intermolecular enthalpy contributions, responsible for the cooperative effect, by some 840 J mol^{-1} on going from the low-spin to the high-spin state [86].

A simple statistical approach has also been tried to account for the metal dilution effect on the spin transition behaviour in $[Fe_xZn_{1-x}(2\text{-pic})_3]Cl_2 \cdot EtOH$ [182]. Here, the spin transition is again considered to take place in domains containing an equal number n of HS and LS molecules, respectively.

$$(LS)_n \rightleftharpoons (HS)_n . \tag{6.1}$$

The symbol ( )$_n$ denotes one mole of domains of any spin state with n mole-cules in each domain. Interactions between the domains, irrespective of the spin state, are neglected. The total energy of an individual complex molecule is consi-dered to be composed of electronic and intramolecular vibrational contributions. In terms of the partition functions for these contributions the equilibrium constant K = $x_{HS}/(1-x_{HS})$ may be formulated as

$$ K = \left\{ \left(\sum_{j=1}^{15} e^{-\epsilon_j/kT} \right) \cdot \exp\left[-\frac{15}{2} \cdot \frac{hc}{kT} (\tilde{\nu}_H - \tilde{\nu}_L) \right] \left[\frac{1-\exp(-hc\,\tilde{\nu}_L/kT)}{1-\exp(hc\,\tilde{\nu}_H/kT)} \right]^{15} \right\}^n \quad (6.2) $$

This expression is based on the approximation that the 15 normal modes of vi-bration of a FeN$_6$ chromophore in each spin state may be reasonably well represented by one "mean effective" vibrational wave number $\tilde{\nu}_L$ and $\tilde{\nu}_H$ for the LS state and the HS state, respectively. The ϵ_j are the energy separations of the 15 electronic levels of the $^5T_2(O_h)$ term, split by crystal field distortion and eventually by spin-orbit coupling, from the $^1A_1$ level. In a reasonable approximation the 15 ϵ_j parameters may be reduced to only two, viz. the energy separation ϵ between the baricenter of the $^5T_2$ term and the $^1A_1$ level, and the trigonal distortion parameter $\delta_t =$ E(^{5}A) – E(^{5}E). An adjustment of the expression

$$ x_{HS}(T) = K/(1 + K), \quad (6.3) $$

after insertion of Eq. (6.2), to the experimental data of $x_{HS}(T)$ is then possible with five parameters, viz. ϵ, δ_t, $\tilde{\nu}_L$, $\tilde{\nu}_H$, and the domain size n. The procedure has been employed to the [Fe$_x$Zn$_{1-x}$(2-pic)$_3$] Cl$_2$ · EtOH system [182]. The domain size n for the pure iron compound was found to be about 6, whereas n was calculated to be about 4 employing the phenomenological thermodynamic model [86]. The average value for the trigonal distortion parameter δ_t of the solid solutions of all x was found to be – 447 cm^{-1}, in good agreement with the value of – 458 cm^{-1} obtained from the analysis of the temperature dependence of the quadrupole splitting of the HS state [85]. ϵ was estimated to be ca. 760 cm^{-1}. The average values found for the mean effective vibrational frequencies $\tilde{\nu}_L$ = 322 cm^{-1} and $\tilde{\nu}_H$ = 282 cm^{-1} compare well with metal-ligand stretching frequencies of the LS and HS phases, respectively, ob-served by FIR spectroscopy on other spin crossover systems [33,77,105].

6.1.2 Influence of Crystal Solvent Molecules

The ground state properties of [Fe(2-pic)$_3$] Cl$_2$ · Sol crystallizing with different solvent molecules "Sol" are found to depend strongly on the nature of the solvent molecule. A Mössbauer effect study [110] has shown that a temperature dependent HS ⇌ LS spin transition occurs in the solvates with Sol = C$_5$H$_5$OH, CH$_3$OH, and H$_2$O, with an increasing stabilization of the $^1A_1$ ground state in the given order (see Fig. 23). The ethanolate has a transition temperature around 121 K, the methanolate one around 150 K. The monohydrate is unique among these solvates in that it has been

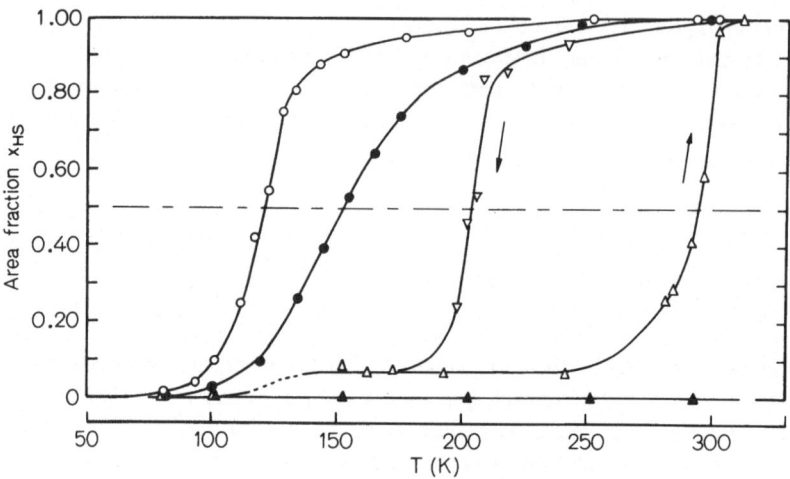

Fig. 23. Temperature dependence of the high-spin area fraction x_{HS} of $[Fe(2\text{-pic})_3]Cl_2 \cdot S$, with $S = C_2H_5OH$ (○), CH_3OH (●), H_2O on heating (△), H_2O on cooling (▽) and $2\,H_2O$ (▲). The high-spin content for the monohydrate is the sum of two high-spin states (A) and (B)

found to exhibit a very large hysteresis (T_c = 204 K for the cooling direction, T_c = 295 K for the heating direction). The dihydrate showed Mössbauer resonances typical of a LS iron(II) species at all temperatures under study [110]. The remarkable influence of the non-coordinating solvent molecule on the spin transition behaviour has been accounted for in terms of a combined action of both the hydrogen-bond formation and changes in the relevant phonon system caused by different packing geometries and/or changes in the crystal structure. A substitution of the intervening solvent molecule is expected to modify the dispersion relations of some normal modes of lattice vibrations as well as the strength of the hydrogen bond. It is obvious, that, if some of these modes respond sensitively to a change in spin state, the spin-crossover characteristics should change with the nature of the solvent molecule.

The above crystal solvent effect on the ethanolate and the methanolate of $[Fe(2\text{-pic})_3]\,Cl_2 \cdot Sol$ has later been confirmed by magnetic susceptibility measurements [181]. Sinn et al. [181] have also investigated the unsolvated bromide $[Fe(2\text{-pic})_3]\,Br_2$ as well as its solvates with $Sol = C_2H_5OH$ and CH_3OH, respectively; these systems all show temperature dependent spin transition with pronounced differences in the transition behaviour. Differences in the spin transition behaviour of corresponding solvates $[Fe(2\text{-pic})_3]\,X_2 \cdot Sol$ (X = Cl, Br) with different anions are also noticeable [181]:

Single crystal X-ray structure analysis has been performed on the ethanolate [31,180,181] the methanolate [94,180] and the dihydrate [94] of $[Fe(2\text{-pic})_3]\,Cl_2$. The space groups found at room temperatue are: $P2_1/c$ (monoclinic with Z = 4) for the ethanolate [31,181], Pbca (orthorhombic with Z = 8) for the methanolate [94], and P1 (triclinic with Z = 2) for the dihydrate [94]. Thus the structures of the three solvates are definitely different, and from this point of view the different magnetic behaviour should not be surprising at all.

The geometry of the $[Fe(2\text{-pic})_3]^{2+}$ cationic complex was found to be the same in the ethanolate [31,180,181] and the methanolate [94,180], viz. meridional (C_{2v}), whereas that in the dihydrate was found to be facial with approximate threefold symmetry [94,180]. A projection of the crystal structure in the low-spin state along the a axis is seen in Fig. 24 a. A perspective drawing of the $[Fe(2\text{-pic})_3]^{2+}$ complex ion in the LS state (90 K) and in the HS state (298 K) is seen in Fig. 24b.

There are also major differences in the hydrogen bond formation as revealed by the crystal structure determinations. Mikami et al. [31] state that in the ethanolate all the amino nitrogen atoms of the $[Fe(2\text{-pic})_3]^{2+}$ complexes are linked to Cl^- ions by N–H ... Cl hydrogen bonds and also the ethanolate molecule is hydrogen bonded to a Cl^- ion by a Cl ... H–O. Katz and Strouse [180] have confirmed this kind of hydrogen bond network in the ethanolate and also for the methanolate. Greenaway and Sinn [194] have reported that, in the dihydrate, hydrogen bonds are formed between the $[Fe(2\text{-pic})_3]^{2+}$ complex and the water molecules, presumably facilitated by the facial geometry of the cationic complex having all three amino nitrogen atoms on one side of the ion. This kind of hydrogen bonding (solvent-to-cation) is absent in the methanolate and in the ethanolate.

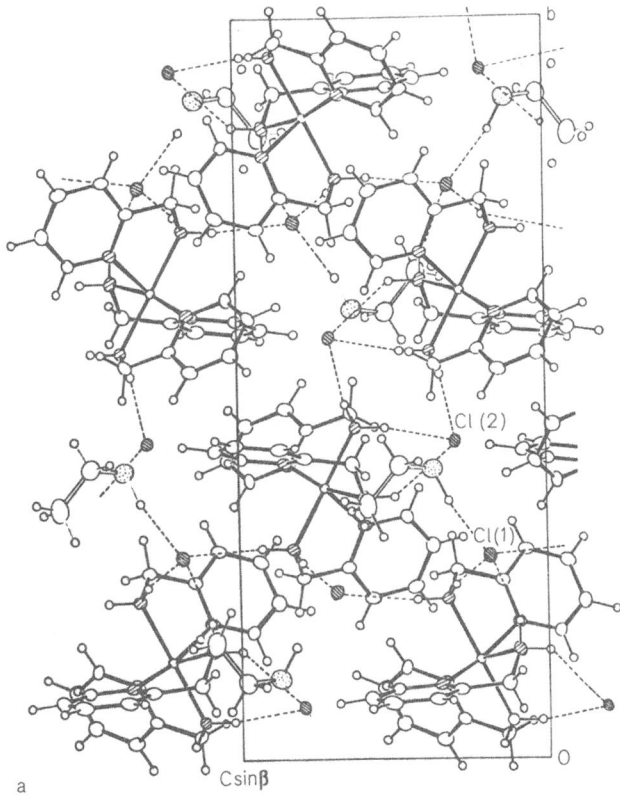

Fig. 24a. Projection of the crystal structure in the low-spin state of $[Fe(2\text{-pic})_3]Cl_2 \cdot EtOH$ along the a-axis (from Ref. 31)

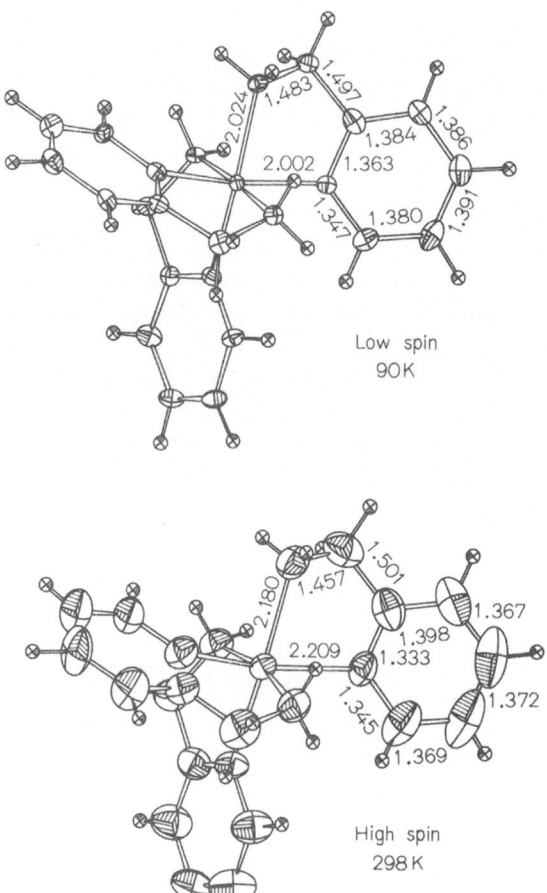

Low spin
90 K

High spin
298 K

b

Fig. 24b. A perspective drawing of the $[Fe(2\text{-pic})_3]^{2+}$ complex ion in the LS state (90 K) and in the HS state (298 K) (from Ref. 31)

The variable-temperature crystal structure analysis of Mikami et al.[31] has also revealed how the metal-ligand distances of the $[Fe(2\text{-pic})_3]^{2+}$ complex change with the spin transition: $Fe-N(NH_2)$ is 2.180 Å at 298 K and 2.024 Å at 98 K, $Fe-N(py)$ is 2.909 Å at 298 K and 2.002 Å at 98 K. Thus the $Fe-N(NH_2)$ changes by 0.156 Å and the $Fe-N(py)$ by 0.207 Å on going from the HS to the LS isomer. The considerably larger decrease in the $Fe-N(py)$ bond is due to the more effective π-backbonding capability in this bond as compared to the $Fe-N(NH_2)$ bond.

Another most interesting phenomenon has been observed by Mikami et al.[31]: The ethanol molecules exhibit orientational disorder over three different sites in the high-spin phase, but orient themselves more and more in one of the three sites with decreasing temperature. The variation with temperature of the orientational ground state population seems to be strongly correlated with the temperature dependent spin transition. The authors therefore suggest that the disorder-order transition of the ethanol molecule "triggers" the spin state transition.

6.1.3 Deuterium Isotope Effect

An interesting H/D isotope effect has recently been observed in both the ethanolate and the methanolate of $[Fe(2\text{-pic})_3] Cl_2$ [131]. The transition temperature T_c in $[Fe(2\text{-pic})_3] Cl_2 \cdot C_2H_5OH$ shifts from 122 K to 136 K, if C_2H_5OH is replaced by C_5H_5OD; a similar shift of T_c from 150 to 165 K occurs in $[Fe(2\text{-pic})_3] Cl_2 \cdot CH_3OH$, if CH_3OH is substituted by CH_3OD (see Fig. 25). This isotope effect may be explained, at least qualitatively, by considering the energy differences ΔE between the zero-point vibronic levels of the HS and the LS state. It can be shown [131] that ΔE

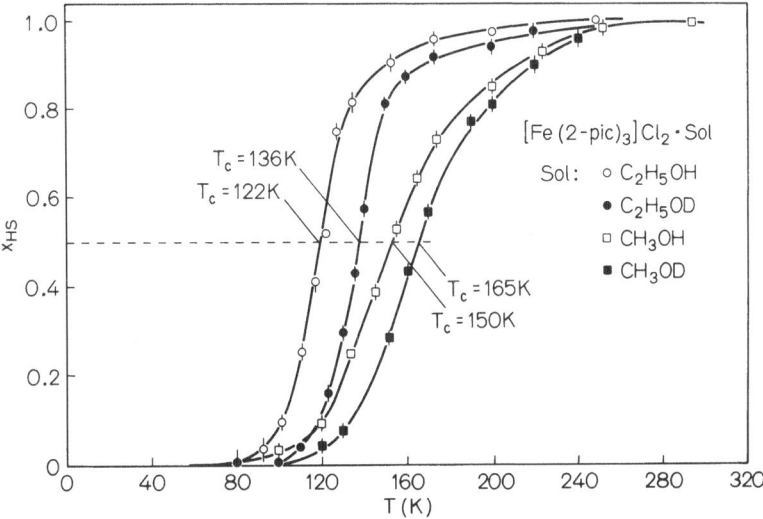

Fig. 25. Temperature dependence of the area fraction x_{HS} of the high-spin quadrupole doublet in the ^{57}Fe Mössbauer spectra of $[Fe(2\text{-pic})_3]Cl_2 \cdot Sol$, Sol = C_2H_5OH (o), C_2H_5OD (•), CH_3OH (•), CH_3OH (□), CH_3OD (■) [131]

changes on going from the deuterated (d) to the non-deuterated (h) solvate, and that the change $\delta(\Delta E) = \Delta E(d) - \Delta E(h)$ may be approximated by

$$\delta(\Delta E) = 15 \cdot \frac{\hbar}{2} (\sqrt{k_{HS}^*} - \sqrt{k_{LS}^*}) \left(\sqrt{\frac{1}{m^*(d)}} - \sqrt{\frac{1}{m^*(h)}} \right) \qquad (6.4)$$

$k_{HS,LS}^*$ are effective force constants, averaged over the 15 valence vibrations of a six-coordinated iron atom, in the HS and the LS state, respectively; they are considered unchanged by deuteration. The reduced mass $m^* = m_{HS}^* = m_{LS}^*$ changes, of course, upon deuteration: $m^*(d) > m^*(h)$. As $k_{HS}^* < k_{LS}^*$, which is known from FIR spectroscopy, $\delta(\Delta E) > 0$, i.e. $\Delta E(d) > \Delta E(h)$, which implies that the LS state is stabilized with respect to the HS state and thus the transition temperature T_c is shifted to higher temperatures upon deuteration.

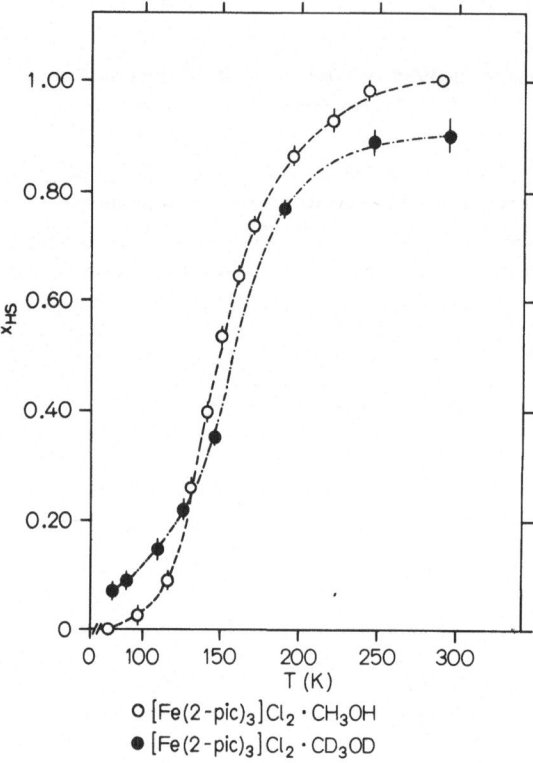

Fig. 26. Temperature dependence of the area fraction x_{HS} of the high-spin quadrupole doublet in the ^{57}Fe Mössbauer spectra of [Fe(2-pic)$_3$]Cl$_2 \cdot$ Sol, Sol = CH$_3$OH (o), CD$_3$OD (•) [132]

o [Fe(2-pic)$_3$]Cl$_2 \cdot$ CH$_3$OH

• [Fe(2-pic)$_3$]Cl$_2 \cdot$ CD$_3$OD

A pronounced H/D isotope effect has also been observed by replacing CH$_3$OH in [Fe(2-pic)$_3$] Cl$_2 \cdot$ CH$_3$OH by fully deuterated CD$_3$OD, as can be seen from Fig. 26 [132].

6.2 Tris(pyridylimidazole) and Related Complexes

Chiswell et al. [133] isolated deep red crystals of [Fe(pyim)$_3$] (ClO$_4$)$_2 \cdot$ H$_2$O, where *pyim* is 2-(2'-pyridyl)imidazole, with a magnetic moment of 5.42 B.M. at 293 K. The same compound, together with a number of other salts of [Fe(pyim)$_3$] X$_2 \cdot$ nH$_2$O, were studied later by Dosser et al. [134]. The magnetic moments of all these complexes fall continuously with temperature (see Table 6); at and above room temperature the moments are still rising. The authors explain this behaviour with the presence of an equilibrium between ^{5}T$_2$(O$_h$) and ^{1}A$_1$(O$_h$) states. The ^{57}Fe Mössbauer spectra, recorded at temperatures down to 4.2 K, readily distinguish

(pyim)

Table 6. Magnetic data of pyridylimidazole and related complexes

Compound	T(K)	μ_{eff} (B.M.)	Ground state[a]	Ref.
$[Fe(pyim)_3]X_2 \cdot nH_2O$				
$X = \frac{1}{2}SO_4^{2-}$ (n = 3)	293/193/113	3.65/2.93/2.59	$^5T_2 \rightleftharpoons {}^1A_1$	134
$X = Cl^-$ (n = 2.5)	293/193/113	2.82/1.39/1.04	$^5T_2 \rightleftharpoons {}^1A_1$	134
$X = BPh_4^-$ (n = 3)	293/193/113	1.62/1.34/113	$^5T_2 \rightleftharpoons {}^1A_1$	134
$X = SCN^-$ (n = 2)	293/193/113	3.05/1.42/1.06	$^5T_2 \rightleftharpoons {}^1A_1$	134
$X = \frac{1}{2}S_2O_3^{2-}$ (n = 4)	293/193/113	2.74/1.83/1.59	$^5T_2 \rightleftharpoons {}^1A_1$	134
$X = ClO_4^-$ (n = 2)	293/193/93	2.78/1.44/0.94	$^5T_2 \rightleftharpoons {}^1A_1$	134
$X = \frac{1}{2}SeO_4^{2-}$ (n = 1)	293/193/113	2.30/1.40/1.24	$^5T_2 \rightleftharpoons {}^1A_1$	134
$X = \frac{1}{2}SO_4^{2-}$ (n = 2)	373/293/193/ 93	4.67/4.08/3.39/ 2.92	$^5T_2 \rightleftharpoons {}^1A_1$	134
$[Fe(6-CH_3-pyim)_3](ClO_4)_2 \cdot H_2O$	298/77	5.14/4.99	$^5T_2$	138
$[Fe(pyim)_2X_2]; X = CN^-$	283/77	0.69/0.42	$^1A_1$	137
$X = Cl^-, Br^-, NCS^-, N_3^-$	283-77	~ 5.2 - ~ 4.9	$^5T_2$	137
$[Fe(pyben)_3]X_2 \cdot nH_2O$				
$X = ClO_4^-$ (n = 1)	306/181/80	5.43/3.97/2.00	$^5T_2 \rightleftharpoons {}^1A_1$	142
$X = ClO_4^-$ (n = 2)	310/190/80	5.35/2.60/1.29	$^5T_2 \rightleftharpoons {}^1A_1$	142
$X = NO_3^-$ (n = 1)	296/180/79	5.35/5.26/4.86	$^5T_2 \rightleftharpoons {}^1A_1$	142
$X = Br^-$ (n = 0)	287/185/80	5.44/5.11/3.65	$^5T_2 \rightleftharpoons {}^1A_1$	142
$X = I^-$ (n = 0)	~ 300/77	~ 5.4/~ 3.6	$^5T_2 \rightleftharpoons {}^1A_1$	143
$X = BPh_4^-$ (n = 1)	~ 300/80	~ 5.0/~ 3.4	$^5T_2 \rightleftharpoons {}^1A_1$	143
$X = BF_4^-$ (n = 0)	294/89	5.38/4.45	$^5T_2 \rightleftharpoons {}^1A_1$	144
$X = BF_4^-$ (n = 0)	295/90	5.43/4.65	$^5T_2 \rightleftharpoons {}^1A_1$	139
$X = BF_4^-$ (n = 1)	~ 300/77	~ 5.4/~ 4.2	$^5T_2 \rightleftharpoons {}^1A_1$	143
$X = BF_4^-$ (n = 2)	~ 300/77	~ 5.0/~ 1.4	$^5T_2 \rightleftharpoons {}^1A_1$	143
$X = BF_4^-$ (n = 3)	353/299/196/ 127/89	5.31/5.04/2.75/ 1.89/1.49	$^5T_2 \rightleftharpoons {}^1A_1$	144
$[Fe(ppp)_3](ClO_4)_2$	290/110	~ 4.4/~ 2.9	$^5T_2 \rightleftharpoons {}^1A_1$	146

[a] Term symbols given in the approximation of O_n symmetry

between the low-spin and the high-spin forms of iron(II), with varying extent of residual paramagnetism depending on the nature of the anion. μ_{eff}, at any given temperature, also varies with the anion. Moreover, the spin crossover behaviour seems also to be influenced markedly by the amount of crystal water in the only case examined (viz. the sulphates). A theoretical calculation of $\chi(T)$, taking a Boltzmann distribution over the $^1A_1$ state and the 15 levels arising from the $^5T_2$ state considering cubic symmetry and spin-orbit interaction, was successful only if the energy gap $\Delta E = \epsilon(^5T_2) - \epsilon(^1A_1)$ was treated as a function of temperature.

Independently of Dosser et al., the perchlorate $[Fe(pyim)_3](ClO_4)_2 \cdot H_2O$ was studied by Goodgame and Machado [135]. Their results (μ_{eff} varying between 3.66 B.M. at 342 K and 0.63 B.M. at 86 K and separate quadrupole doublets for the two spin states) agree well with those of Dosser et al., although the latter authors reported on a perchlorate complex with two water molecules per unit formula. An attempt by Goodgame and Machado to describe the temperature dependence of the magnetic susceptibility by a simple Boltzmann distribution assuming a fixed energy gap $\Delta E = \epsilon(^5T_2) - \epsilon(^1A_1)$ failed. From a plot of lnK vs. 1/T they estimated an enthalpy change of $\Delta H = 4.2$ kcal mol^{-1} and an entropy change of $\Delta S = 12.3$ eu over the range 340–225 K. As the entropy change due to a simple change of spin state, from singlet to quintet, is only Rln 5 = 3.2 eu, the authors suggested that additional factors such as changes in bond lengths and angles play an important part in the spin transition phenomenon.

Goodgame and Machado have also studied the magnetism of two forms (A and B) of the complex [Fe(pyimi)$_3$] (ClO$_4$)$_2$, where *pyimi* is 2-(2'-pyridyl)imidazoline,

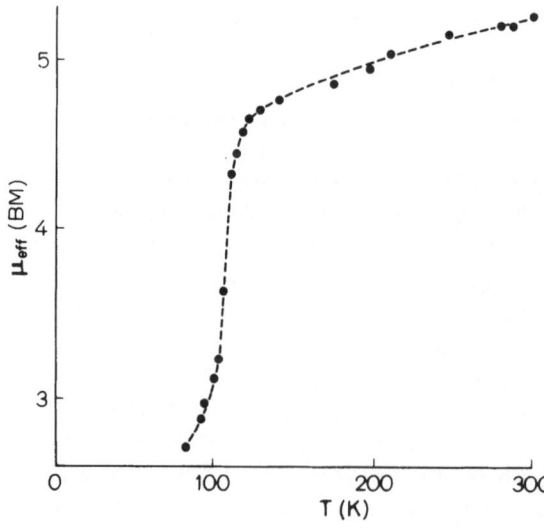

(pyimi)

which they prepared by two different methods[136]. They found that the dark blue form A is low-spin with a magnetic moment and with Mössbauer parameters general-ly expected for $^1A_1$ states of iron(II). The dark purple form B differs not only in the X-ray powder pattern from form A, but also in the magnetism. The magnetic moment indicates high-spin behaviour at room temperature ($\mu_{eff} = 5.25$ B.M. at 295 K), but drops sharply around 120 K to 2.7 B.M. at 83 K (see Fig. 27). The susceptibility $\chi(T)$ was observed to be time-dependent near the transition tempe-rature.

The replacement of one *pyim* ligand by two monodentate anions X, resulting into bis-complexes of the type [Fe(pyim)$_2$X$_2$], "destroys" the spin crossover phenomenon[137]: The complexes with X = Cl$^-$, Br$^-$, NCS$^-$, N$_3^-$ are all high-spin; the complex with X = CN$^-$ is low-spin. The same authors have investigated the effect of introducing a methyl group in 6-position of the *pyim* ligand to yield

Fig. 27. Temperature dependence of the effective magnetic moment μ_{eff} of [Fe(pyimi)$_3$](ClO$_4$)$_2$, form B (from Ref. 136)

$[Fe(6-CH_3-pyim)_3](ClO_4)_2 \cdot H_2O$ [138]. Due to extensive steric hindrance, causing the metal-ligand bond to lengthen and weaken, this complex is high-spin over the whole temperature range studied; some magnetic data are listed in Table 6.

Enhanced steric hindrance with the tendency of relative stabilization of the $^5T_2$ state is also expected in the $[Fe(pyben)_3]^{2+}$ complex ion, with *pyben* being 2-(2'-pyridyl)benzimidazole, as compared to the unsubstituted pyim complex. Several research groups have reported on magnetic studies of $[Fe(pyben)_3]X_2 \cdot nH_2O$ complexes [138–143].

(pyben)

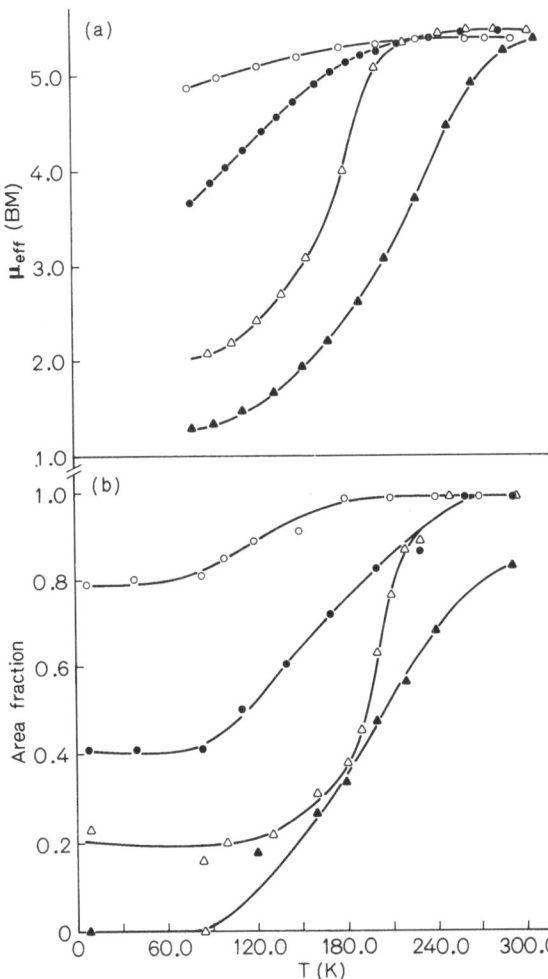

Fig. 28a, b. Temperature dependence of (a) the effective magnetic moment μ_{eff} and (b) the Mössbauer area fraction of the high-spin quadrupole doublet in $[Fe(pyben)_3]$-$(ClO_4)_2 \cdot 2 H_2O$ (▲); $[Fe(pyben)_3](ClO_4)_2 \cdot H_2O$ (△); $[Fe(pyben)_3]Br_2$ (●); $[Fe(pyben)_3](NO_3)_2 \cdot H_2O$ (○) (from Ref. 142)

All of these salts have been described as spin crossover systems with more or less completion of the spin conversion. The perchlorate shows the most complete conversion between room temperature and $\sim$ 100 K (see Fig. 28a). The magnetic moments of the bromide and, particularly, the nitrate vary to an extent which is much less pronounced than in the perchlorate (see Fig. 28a), and it would be difficult to characterize the ground state behaviour on the magnetic moments alone. The Mössbauer spectra, however, exhibit clearly the coexistence of the $^5T_2(O_h)$ and $^1A_1(O_h)$ states in all cases. As an example, Fig. 29 shows the bromide spectrum at 80 K. The temperature dependence of the Mössbauer area fraction of the high-spin quadrupole doublet parallels the temperature variation of μ_{eff} in all cases shown in Fig. 28a, b. There is a distinct anion dependence of μ_{eff} at any given temperature[139,140], but no obvious correlation between μ_{eff} and the size of the anion is found. Also, the amount of crystal water seems to exert a certain influence as can be seen from the magnetic data in Table 6 as well as from Fig. 28. This has been rationalized with differences in hydrogen bonding between water molecules and the imino-hydrogen atom of the ligand. Comparison with the *pyim* complexes suggests that the $^5T_2(O_h)$ manifold is thermally accessible at considerable lower temperatures for the $[Fe(pyben)_3]^{2+}$ salts; this agrees with prediction. From magnetically perturbed Mössbauer spectra, Sams et al.[142] have deduced that the $[Fe(pyben)_3]^{2+}$ complex ion adopts the meridional-isomer configuration in both spin states.

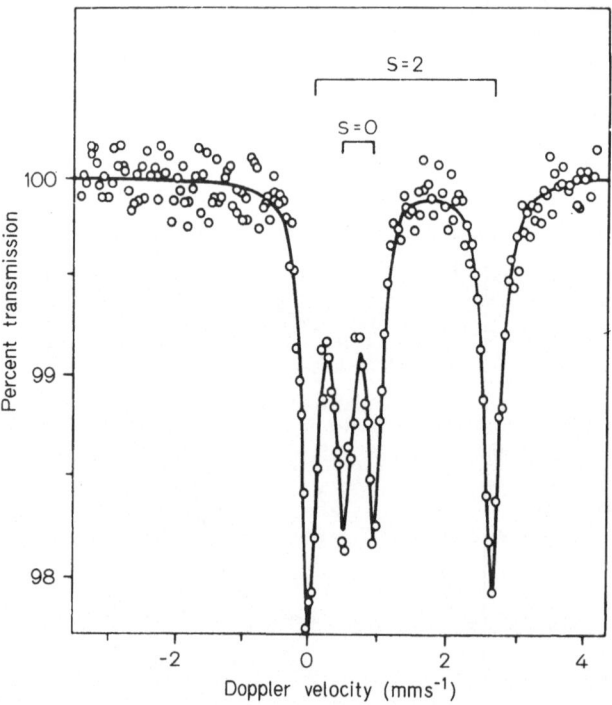

Fig. 29. Mössbauer spectrum of $[Fe(pyben)_3]Br_2$ at 80 K (from Ref. 140)

Temperature-dependent spin conversion, to a different extent in the range 300–80 K, occurs also in the [Fe(pyben)$_3$] X$_2$ · nH$_2$O salts with X = I$^-$ [139,143], BF$_4^-$ [139] both in the monohydrate and the dihydrate [143], X = NCS$^-$ [143] X = BPh$_4^-$ [143], and X = [Cr(NH$_3$)$_2$(NCS)$_4$]$^-$ [143]. Some representative magnetic data are given in Table 6. In all these compounds, even in those with relatively high magnetic moments in the low-temperature region, the coexistence of $^5T_2$(O$_h$) and $^1A_1$(O$_h$) states has been established by Mössbauer spectroscopy, with the relative intensities of the high-spin and low-spin quadrupole doublets changing with temperature to variable extents. The area fraction of the $^5T_2$ doublet is 0.40 (4.2 K) for X = I$^-$, 0.89 (4.2 K) for X = NCS$^-$ (n = 1), 0.52 (4.2 K) for X = BF$_4^-$ (n = 1), 0.0 (115 K) for X = BF$_4^-$ (n = 2), 0.33 (4.2 K) for X = BPh$_4^-$ (n = 1), 0.0 (115 K) for X = [Cr(NH$_3$)$_2$(NCS)$_4$]. The tetrafluoroborate complexes with one and two water molecules per unit formula differ considerably in their crossover behaviour, even more than the monohydrate and the dihydrate of the perchlorate. Again, differences in hydrogen bond formation between the acidic amino-hydrogen atom and water molecules, possibly including the anion as an intervening group similar to the Cl$^-$ anion in [Fe(2-pic)$_3$] Cl$_2$ · EtOH [31], have been considered responsible for this effect. It has been postulated that this hydrogen bond formation increases the acidity of the NH group of the ligand, which in turn strengthens the nitrogen-iron bond through enhanced σ-donation, thereby favouring the low-spin state.

Replacing one *pyben* ligand by two NCS$^-$ ligands yields the complex [Fe(pyben)$_2$(NCS)$_2$], which remains fully high-spin down to 4.2 K. This is somewhat surprising, at first glance, because the electronic spectra [142] indicate that 10 Dq is slightly larger in the high-spin complex [Fe(pyben)$_2$(NSC)$_2$] than in [Fe(pyben)$_3$] (NCS)$_2$ · H$_2$O, which shows spin crossover. The authors suggested, that hydrogen bond formation, which is only possible in the latter system, may be the reason for this contradiction.

The significant dependence of the magnetic properties of [Fe(pyben)$_3$] (BF$_4$)$_2$ · nH$_2$O on the extent of hydration has been further persued by Goodwin et al. [144], who have examined the trihydrate and the anhydrous complex. The magnetic moment of both forms decreases on cooling, indicating a gradual increase in the proportion of low-spin species present; the change is much more pronounced for the trihydrate than for the anhydrous complex. In fact, inspection of all four forms of [Fe(pyben)$_3$] (BF$_4$)$_2$ · nH$_2$O with n = 0, 1, 2, 3 shows a correlation with their magnetic moments at comparable temperatures, viz. the low-spin state is favoured with increasing number of lattice water molecules. The more lattice water present, the more facilitated will be hydrogen bond formation, which in turn will be the cause for increasing acidity of the NH group and along with it the increased strengthening of the metal-ligand σ-bond.

Sams and Tsin [142,143] stated, that the spin crossover phenomenon in [Fe(pyim)$_3$]$^{2+}$ and [Fe(pyben)$_3$]$^{2+}$ complexes probably occurs only in the solid state implying that the phenomenon should depend not only on the field strength of the ligands coordinated to iron, but is ultimately controlled by lattice forces in the crystal. Contrary to this statement, Reeder, Dose and Wilson [145] proved by variable-temperature magnetic and electronic spectral studies on solutions of the tetraphenylborates of [Fe(pyim)$_3$]$^{2+}$ and [Fe(pyben)$_3$]$^{2+}$ in CH$_3$CN (20%)/CH$_3$OH and acetone, respectively, that a spin equilibrium $^5T_2$(O$_h$) ⇌ $^1A_1$(O$_h$) exists also in solution. The magnetic moments, determined by the Evans ^{1}H-NMR method, are plotted as a function of temperature in Fig. 30. From

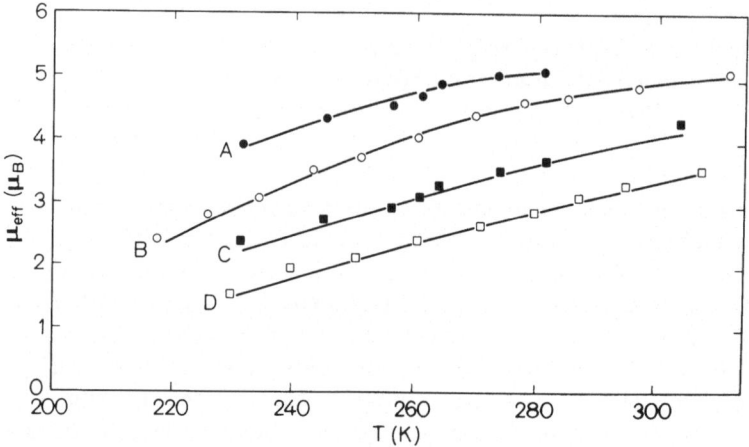

Fig. 30. Temperature dependence of the effective magnetic moment μ_{eff} in solution for (A) [Fe(pyben)$_3$](BPh$_4$)$_2$ in CH$_3$CN (20%)/CH$_3$OH, (B) [Fe(pyben)$_3$](BPh$_4$)$_2$ in acetone, (C) [Fe(pyim)$_3$](BPh$_4$)$_2$ in CH$_3$CN (20%)/CH$_3$OH, (D) [Fe(pyim)$_3$](BPh$_4$)$_2$ in acetone (from Ref. 145)

this graph it appears that the high-spin ($^5T_2$) state, at any given set of temperature/solvent/anion conditions, is more stable in the [Fe(pyben)$_3$]$^{2+}$ complex than in the [Fe(pyim)$_3$]$^{2+}$ complex. This implies that 10 Dq(pyim) > 10 Dq(pyben), which seems reasonable because of the larger steric requirements of the *pyben* ligand producing the longer (and weaker) iron-nitrogen bonds and thus the smaller 10 Dq value as compared to the *pyim* ligand with less steric hindrance. The temperature dependent spin state equilibrium in solution is also apparent from Fig. 31, where the high-intensity "low-spin band" (with extinction coefficients ranging from 1000 to 5000) at 486 nm, most likely being charge transfer in origin, is plotted as a function of temperature.

The ligand 3-phenyl-5-pyridyl-(2)-pyrazole (ppp) yields iron(II) complexes, which are structurally closely related to the [Fe(pyim)$_3$]$^{2+}$ and [Fe(pyben)$_3$]$^{2+}$ complexes.

(ppp)

The magnetic properties also resemble each other. Similarly to the corresponding pyim and pyben complexes, the [Fe(ppp)$_3$](ClO$_4$)$_2$ shows a gradual spin crossover, with magnetic moments ranging from ~ 2.9 at 110 K to ~ 4.4 B.M. at 290 K [146].

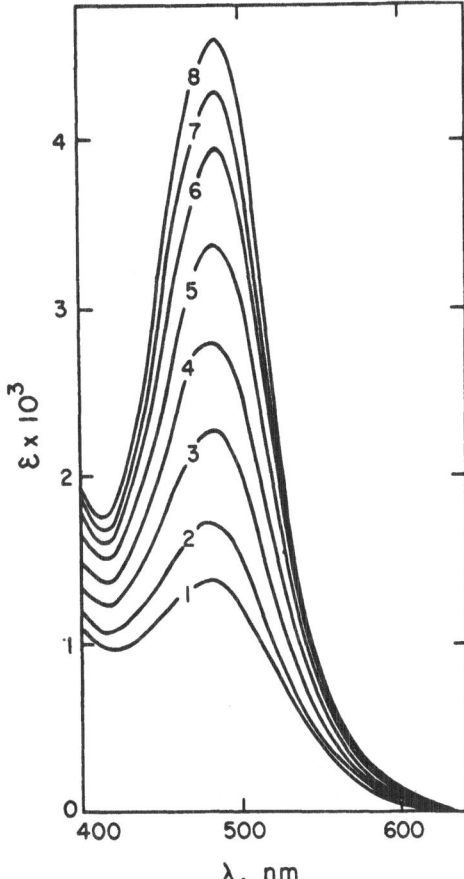

Fig. 31. Variable-temperature electronic spectrum of [Fe(pyim)$_3$](BPh$_4$)$_2$ in CH$_3$CN (20%)/CH$_3$OH at (1) 302 K, (2) 296 K, (3) 285 K, (4) 207 K, (5) 265 K, (6) 254 K, (7) 246 K and (8) 235 K (from Ref. 145)

6.3 2-Pyridinalcarbaldehyde Complexes

Barth, Schmauss, and Specker [147], have reported on the preparation and magnetic studies of a series of substituted tris(2-pyridinalphenylimine) complexes of iron(II). Complexes with fully low-spin ($^1A_1$) and fully high-spin ($^5T_2$) ground states could be prepared by appropriate substitution in 2- and 4-position of the phenyl ring of the 2-pyridinalphenylimine ligand. The only complex showing a distinct decrease of μ_{eff} on cooling is [Fe(ppi)$_2$(NCS)$_2$], where *ppi* is the unsubstituted 2-pyridinal-phenylimine. μ_{eff} decreases only from 4.5 B.M. at 293 K to $\sim$ 3.3 B.M. at 4 K. The

(ppi)

authors ascribed this relatively small change to a reversible transition from $^5T_2(O_h)$ to $^3T_1(O_h)$. Later Mössbauer effect measurements, however, have clearly demonstrated the presence of an incomplete $^5T_2(O_h) \rightleftharpoons {}^1A_1(O_h)$ transition, with ca. 60% of the molecules remaining in the high-spin state at low temperatures and causing the unusually high saturation value of μ_{eff} in the low-temperature region [148].

Maeda, Takashima, and Nishida [149] have prepared iron(II) complexes of pyridincarbaldehyde and quinolinecarbaldehyde ligands. Only complexes of the type $[Fe((2\text{-pym})\text{-R})_2 \, (NCS)_2]$, with 2-pym being 2-pyridylmethylene, and R = isoproylamine, p-anisidine, p-toluidine, aniline, p-chloroaniline, o-toluidine ($\cdot 1/2\,CHCl_3$), o-chloroaniline, were found by Mössbauer spectroscopy and magnetic susceptibility

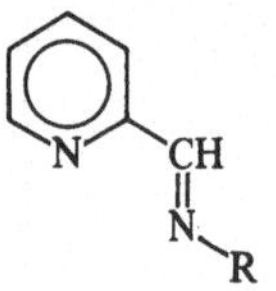

(2-pym)-R

measurements to exhibit an incomplete $^5T_2 \rightleftharpoons {}^1A_1$ transition in the solid. The Mössbauer quadrupole doublets of the high-spin and the low-spin state of iron(II), respectively, are well separated. The van Vleck approach could not be used to account for the magnetism.

7 Complexes with Tridentate Substituted-Pyridine Ligands

The ligands in this class of compounds can be considered to be derived from 2,2',2''-terpyridyl *(terpy)* by replacing either the inner pyridine ring or one or both of the outer pyridine rings by different groups with nitrogen or different donor atoms. In

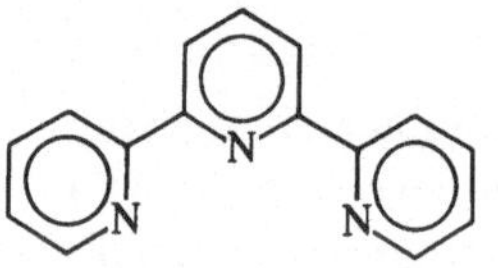

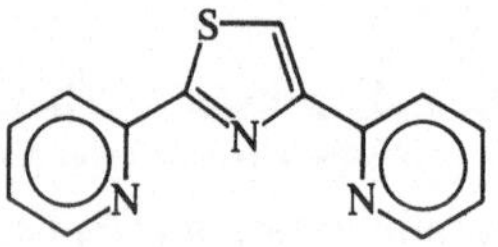

the following we shall subdivide this class of bis(tridentate ligand) complexes of iron(II) according to the core notation, in which we shall denote the pyridine nitrogen atom by N and inequivalent ones by N'.

7.1 Complexes with Fe(NN'N)₂ Core

The most prominent spin crossover systems belonging to this family are the $[Fe(pythiaz)_2]X_2$ and the $[Fe(paptH)_2]X_2$ complexes, were *pythiaz* refers to 2,4-bis-(2-pyridyl)thiazole and *paptH* to 2-(2-pyridylamino)-4-(2-pyridyl)thiazole. Goodwin

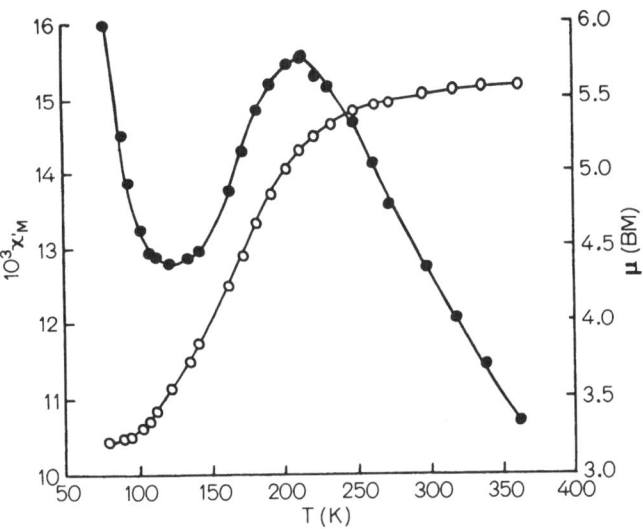

(paptH)

[150)] first reported on well defined chelates of *pythiaz* and *paptH* with iron(II), cobalt(II), and nickel(II). He described the bis-pythiaz complexes to be less stable, apparently due to greater distortion of its bond angles induced on coordination, than the corresponding bis-paptH complexes. Both bases were characterized to yield complexes of less stability than the corresponding complexes of terpyridine. Room temperature magnetic moments indicated that the magnetism of the $[Fe(paptH)_2]$-$X_2 \cdot nH_2O$ complexes varies strongly with the nature of the anion: the complexes with $X = ClO_4^-$ (n = 1), I^- (n = 1) and $[Fe(papt)_2]$ are high-spin with magnetic moments around 5.0 B.M., and the complexes with $X = Cl^-$ (n = 2), Br^- (n = 2) are low-spin with moments of 1.7 and 1.3 B.M., respectively, at ambient temperature. Five-coordinate complexes $[M(paptH)X_2]$ and $[M^{(II)}(pythiaz)X_2]$ (M = Fe, Co, Ni) have also been described; these are all spin-free (paramagnetic) compounds [150,151].

7.1.1 "Pythiaz"-Complexes

A more refined investigation of the magnetic properties of $[Fe(pythiaz)_2]X_2$ ($X = ClO_4^-$, BF_4^-) at temperatures down to 80 K was carried out by Goodwin and Sylva[151]. They found that μ_{eff} decreases gradually from ~ 5.4–5.5 B.M. at 298 K to 3.2–3.4 B.M. at 80 K (see Table 7 and Fig. 32), and suggested that the large residual paramagnetism,

Fig. 32. Temperature dependence of the magnetic susceptibility $\chi(T)$ (●) and the effective magnetic moment μ_{eff} (○) of $[Fe(pythiaz)_2](BF_4)_2$ (from Ref. 151)

Table 7. Magnetic data of iron(II) complexes with tridentate (substituted-pyridine) ligands

Compound	T(K)	μ_{eff} (B.M.)	Ground state [a]	Ref.
$[Fe(pythiaz)_2]X_2; X = ClO_4^-$	298/160/80	5.34/4.56/ 3.45	$^5T_2 \rightleftharpoons {}^1A_1$	151
$X = BF_4^-$	363/298/203/ 123/80	5.59/5.53/ 5.03/3.57/ 3.21	$^5T_2 \rightleftharpoons {}^1A_1$	151
$[Fe(6-CH_3-pythiaz)_2]X_2 \cdot 2H_2O$				
$X = ClO_4^-$	293/83	5.13/4.83	$^5T_2$	152
$X = BF_4^-$	293/83	4.97/4.78	$^5T_2$	152
$[Fe(paptH)_2]X_2 \cdot nH_2O$				
$X = ClO_4^-, n = 1$	306/99	5.14/4.64	$^5T_2$	109
$X = I^-, n = 1$	333/102	5.05/4.77	$^5T_2$	109
$X = Cl^-, n = 0$ (yellow)	300/98	5.39/5.20	$^5T_2$	109
$X = \frac{1}{2}C_2O_4^{2-}, n = 1$	296/86	5.28/4.99	$^5T_2$	109
$X = \frac{1}{2}SO_4^{2-}, n = 3.5$	302/98	5.08/4.80	$^5T_2$	109
$X = SCN^-, n = 1$	339/245/136	4.82/3.58/ 2.25	$^5T_2 \rightleftharpoons {}^1A_1$	109
$X = NO_3^-, n = 1$	357/254/163	4.12/1.68/ 0.97	$^5T_2 \rightleftharpoons {}^1A_1$	109
$X = Cl^-, n = 2$	380/280/178	3.94/1.60/ 0.69	$^5T_2 \rightleftharpoons {}^1A_1$	109
$X = Cl^-, n = 0$ (red-brown)	347/242/166	2.84/1.04/ 0.68	$^5T_2 \rightleftharpoons {}^1A_1$	109
$X = Br^-, n = 2$	404/364/283	3.49/2.58/ 1.01	$^5T_2 \rightleftharpoons {}^1A_1$	109
$X = \frac{1}{2}SO_4^{2-}, n = 5$	384/266/127	4.60/2.57/ 1.44	$^5T_2 \rightleftharpoons {}^1A_1$	109
$X = \frac{1}{2}PtCl_6^{2-}, n = 2$	384/245/106	4.08/2.84/ 1.76	$^5T_2 \rightleftharpoons {}^1A_1$	109
$[Fe(papt)_2] \cdot Sol \quad Sol = 0$	310/190/94	4.96/3.35/ 1.55	$^5T_2 \rightleftharpoons {}^1A_1$	155
$Sol = C_6H_6$	298/94	4.92/4.29	$^5T_2 (\rightleftharpoons {}^1A_1)$	155
$Sol = 4/3 \; CHCl_3$	297/96	5.02/4.58	$^5T_2 (\rightleftharpoons {}^1A_1)$	155
$[Fe(Y-paptH)_2]X_2 \cdot nH_2O$				
$Y = 3-CH_3, X = Cl^- (n = 0)$	293/83	5.26/5.19	$^5T_2$	152
$Y = 3-CH_3, X = Br^- (n = 3)$	303/173/83	5.04/3.69/ 3.22	$^5T_2 \rightleftharpoons {}^1A_1$	152
$Y = 3-CH_3, X = I^- (n = 2)$	303/173/83	5.10/4.76/ 4.36	$^5T_2 \rightleftharpoons {}^1A_1$	152
$Y = 3-CH_3, X = NO_3^- (n = 2)$	303/183/83	5.09/4.42/ 3.63	$^5T_2 \rightleftharpoons {}^1A_1$	152
$Y = 3-CH_3, X = BF_4^- (n = 1)$	303/183/83	5.21/4.56/ 4.07	$^5T_2 \rightleftharpoons {}^1A_1$	152
$Y = 4-CH_3, X = Cl^- (n = 3)$ (red-brown form)	303/203/83	1.24/0.48/ 0.43	$^1A_1$	152
$Y = 4-CH_3, X = Cl^- (n = 3)$ (brown form)	343/203/83	3.64/2.09/ 1.31	$^5T_2 \rightleftharpoons {}^1A_1$	152
$Y = 4-CH_3, X = Br^- (n = 3)$	343/243/83	2.13/0.95/ 0.55	$(^5T_2 \rightleftharpoons {}^1A_1)$	152

Table 7 (continued)

Compound	T(K)	μ_{eff} (B.M.)	Ground state[a]	Ref.
Y = 4–CH$_3$, X = I$^-$ (n = 2)	293/173/83	1.96/1.03/ 0.90	($^5T_2 \rightleftharpoons {}^1A_1$)	152
Y = 4–CH$_3$, X = $\frac{1}{2}$ SO$_4^{2-}$ (n = 3)	293/193/83	3.65/2.67/ 2.17	$^5T_2 \rightleftharpoons {}^1A_1$	152
Y = 4–CH$_3$, X = BF$_4^-$ (n = 2)	313/183/83	4.70/2.57/ 1.56	$^5T_2 \rightleftharpoons {}^1A_1$	152
Y = 4–CH$_3$, X = NO$_3^-$ (n = 2)	323/173/83	2.37/0.79/ 0.52	$^5T_2 \rightleftharpoons {}^1A_1$	152
[Fe(Y-papt)$_2$]·Sol				
Y = 3–CH$_3$, Sol =	343/123/83	4.91/4.71/ 4.60	$^5T_2 (\rightleftharpoons {}^1A_1)$?	152
Y = 4–CH$_3$, Sol =	373/173/83	4.73/3.46/ 2.96	$^5T_2 \rightleftharpoons {}^1A_1$	152
Y = 4–CH$_3$, Sol = 2C$_6$H$_6$	303/203/83	4.68/1.92/ 0.88	$^5T_2 \rightleftharpoons {}^1A_1$	152
Y = 4–CH$_3$, Sol = CHCl$_3$	303/173/83	4.96/4.00/ 3.61	$^5T_2 \rightleftharpoons {}^1A_1$	152
Y = 6–CH$_3$, Sol = CHCl$_3$	303/143/83	4.96/4.78/ 4.62	($^5T_2 \rightleftharpoons {}^1A_1$)	152
[Fe(dbtp)$_2$](ClO$_4$)$_2$·3H$_2$O	313/213/123/ 83	4.49/3.32/ 2.34/2.23	$^5T_2 \rightleftharpoons {}^1A_1$	164

[a]   Term symbols given in the approximation of O$_h$ symmetry

giving rise to magnetic moments around 3.5 B.M. in the low-temperature limit, is possibly due to only 50% conversion of the molecules from $^5T_2(O_h)$ to $^1A_1(O_h)$. The variation of μ_{eff} with temperature is accompanied by a change in colour, from bright red at room temperature to intense purple at about 100 K, closely resembling the colour of the diamagnetic [Fe(terpy)$_2$]$^{2+}$ salts. The decrease of the field strength of pythiaz as compared to terpy is probably caused by the reduced basicity of the nitrogen donor atom of the thiazole rising relative to that of the pyridine ring and, to some extent, by more distorted bond angles induced on coordination of the *pythiaz* ligand.

Introducing a methyl group in α-position next to the nitrogen atom in *pythiaz*, to yield *6–CH$_3$-pythiaz*, causes a considerable decrease in the ligand field strength of the complexes [Fe(6–CH$_3$-pythiaz)$_2$]X$_2$·2H$_2$O (X = ClO$_4^-$, BF$_4^-$), due to the enhanced steric requirements leading to an increase in metal-ligand bond lengths, as compared to the unsubstituted *pythiaz* ligand. Both the perchlorate and the tetrafluoroborate complexes of *6–CH$_3$-pythiaz* are fully high-spin [152].

6–CH$_3$-pythiaz

Measurements of the ^{57}Fe Mössbauer effect in [Fe(pythiaz)$_2$]X$_2$ (X = ClO$_4^-$, BF$_4^-$)[153] have clearly revealed that the continuous decrease of μ_{eff} with temperature arises from a gradual $^5T_2(O_h) \rightleftharpoons {}^1A_1(O_h)$ transition with very high fractions of residual paramagnetism at low temperatures, as was originally suggested [151]. The spectra show two well resolved quadrupole doublets at all temperatures, with parameters typical of iron(II) in the low-spin state and in the high-spin state, respectively. The residual paramagnetism, derived from the low-temperature area fractions of the high-spin quadrupole doublets, amounts to 41–46%. The z-component of the EFG tensor has been determined in an applied magnetic field to be $V_{zz}(^1A_1) < 0$; the determination of $V_{zz}(^5T_2)$ was not successful. From an Ingalls type treatment [154] of the temperature dependence of $\Delta E_Q(^5T_2)$, the axial ligand field splitting in both salts was estimated to range from ca. -500 cm^{-1} (298 K) to -460 cm^{-1} (4.2 K) assuming $V_{zz}(^5T_2) > 0$, and from ca. 690 cm^{-1} (298 K) to ca. 850 cm^{-1} (4.2 K) assuming $V_{zz}(^5T_2) < 0$, respectively. A semiempirical expression for $\mu_{eff}(T)$, taking the fraction α of residual paramagnetism and t.i.p. into account, could be fitted to experimental μ_{eff} data only if a strong temperature dependence of $\epsilon = E(^5T_2) - E(^1A_1)$ was accepted. This has led the authors to conclude that the nature of the transition cannot be interpreted in terms of a "pure" Boltzmann distribution over a fixed set of energy levels.

7.1.2 "Papt(H)" Complexes

A detailed magnetic study in the temperature range 100–350 K of a series of [Fe(paptH)$_2$]X$_2 \cdot n$H$_2$O complexes was performed by Sylva and Goodwin [109]. They found the magnetism to be markedly dependent on the nature of the anion and, in some instances, on the presence of water in the crystal lattice. On the basis of their results, two classes of [Fe(paptH)$_2$]X$_2 \cdot n$H$_2$O complexes may be distinguished: (a) Complexes with X = $\frac{1}{2}$C$_2$O$_4^{2-}$ (n = 1), ClO$_4^-$ (n = 1), $\frac{1}{2}$SO$_4^{2-}$ (n = 3.5), I$^-$ (n = 1), Cl$^-$ (yellow) are paramagnetic and follow the Curie-Weiss law; (b) complexes with X = SCN$^-$ (n = 1), $\frac{1}{2}$PtCl$_6^{2-}$ (n = 2), $\frac{1}{2}$SO$_4^{2-}$ (n = 5), NO$_3^-$ (n = 1), Cl$^-$ (n = 2), Cl$^-$ (n = 0, red-brown), Br$^-$ (n = 2) show temperature dependent gradual spin-crossover (see Table 7 and Fig. 33). A correlation between the magnetism and the colour of the compounds has been established, the high-spin form appearing yellow (orange) and the low-spin form preferentially brown. The magnetic properties of the anhydrous nitrate are particularly unusual. Over the temperature range 383 – 228 K the compound appears to obey the Curie-Weiss law. At T $\sim$ 220 K, however, deviations from the Curie-Weiss behaviour begin to appear and the magnetic moment, at a given temperature, decreases with time accompanied by a change in colour from yellow to brown. Absorption of water by the anhydrous nitrate (yellow, high-spin) is also accompanied by a pronounced colour change, due to the formation of the hydrated nitrate (brown, predominantly low-spin). The authors have ascribed the differences in the magnetic behaviour to differences in crystal lattice forces. This statement receives support from the different shapes of the lnK vs. 1/T plots (equilibrium constant K expressed in terms of mole fractions of the two spin states as determined from the μ_{eff} data), which are all, except for the bromide, not straight lines; this indicates

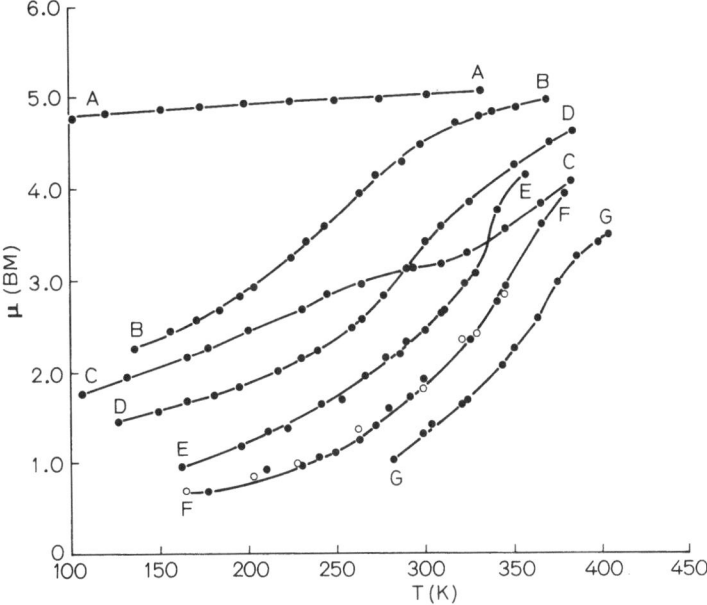

Fig. 33. Temperature dependence of the effective magnetic moment of (A) [Fe(paptH)$_2$]-
I$_2$·H$_2$O; (B) [Fe(paptH)$_2$](SCN)$_2$·H$_2$O; (C) [Fe(paptH)$_2$](PtCl$_6$)·H$_2$O; (D) [Fe(paptH)$_2$]-
SO$_4$·5 H$_2$O; (E) [Fe(paptH)$_2$](NO$_3$)$_2$·H$_2$O; (F) [Fe(paptH)$_2$]Cl$_2$·2 H$_2$O (•) and [Fe(paptH)$_2$]-
Cl$_2$ (red-brown form) (○); (G) [Fe(paptH)$_2$]Br$_2$·2 H$_2$O (from Ref. 109)

that ΔH changes with temperature which rules out that a simple intramolecular
equilibrium involving HS and LS states is occurring. The macroscopic change of ΔH
around 3–5 kcal mol^{-1} undoubtedly arises mainly from the necessary increase of the
intramolecular volume accompanying the change from low-spin to high-spin [31,94].

The neutral complex [Fe(papt)$_2$] (with the acidic imino hydrogen in paptH
stripped off) has also been described as showing temperature dependent spin transition [155], with magnetic moments ranging from 4.96 B.M. (at 310 K) to 1.55 B.M.
(at 94 K). Its well defined solvates, [Fe(papt)$_2$]·C$_6$H$_6$ and [Fe(papt)$_2$]·4/3 CHCl$_3$,
however, have strikingly different temperature dependences of the magnetic moment [155], which decreases only slightly on cooling and reaches a limit of
$\sim$ 4.3–4.6 B.M. at $\sim$ 95 K. It has been suggested that appreaciable spin-pairing in
the solvates causes the slight decrease of the magnetic moment below the values
generally encountered for pure iron(II) high-spin compounds. The striking difference
in the magnetism of [Fe(papt)$_2$] on the one hand, and that of the solvates on the
other, was attributed to differences in crystal lattice forces induced on incorporating
the solvent molecules.

^{57}Fe Mössbauer effect measurements (between 343 and 4.2 K) on [Fe(papt)$_2$]
and its solvates with C$_6$H$_6$ and 4/3 CHCl$_3$ were performed by König, Ritter, and
Goodwin [156]. The Mössbauer spectra, containing two quadrupole doublets with

temperature dependent intensities and parameters characteristic of iron(II) low-spin and high-spin, respectively, provided clear evidence for an incomplete $^5T_2(O_h) \rightleftharpoons$ $^1A_1(O_h)$ transition with very different fractions of permanent $^5T_2(O_h)$ molecules. The area fractions of the high-spin quadrupole doublet, $A(^5T_2)$, found at 4.2 K, were 0.52 for $[Fe(papt)_2] \cdot 4/3$ $CHCl_3$ and 0.81 for $[Fe(papt)_2] \cdot C_6H_6$. This explains the relatively high magnetic moments of $\sim 4.3-4.6$ B.M. observed at ~ 95 K [155]. A reinvestigation of the magnetism of the three systems [156] and a comparison with the results of ref. [155] yielded noticeable differences in the magnetic data. It is therefore very likely that minor differences in the crystal lattice, e.g. in the grain size distribution, disorder and crystal imperfections, exert a decisive influence on the magnetic properties. This is further supported by the observation that two different samples of the unsolvated $[Fe(papt)_2]$ complex, prepared by the same method [155], showed substantial differences in the magnetic moment, particularly at 83 K (3.13 B.M. for sample A, 2.13 B.M. for sample B), as well as in the area fraction $A(^5T_2)$ at 4.2 K (0.43 for sample A, 0.17 for sample B), although the Mössbauer hyperfine parameters (ΔE_Q and δ) were equal within the error limits. With sample B in an applied magnetic field, $V_{zz}(^1A_1) < 0$ and the asymmetry parameter $\eta \sim 0.7$ have been determined; for the $^5T_2$ ground state $V_{zz}(^5T_2) > 0$ and $\eta \sim 0.75$ have been found. The Debye-Waller factors $f(^5T_2)$ and $f(^1A_1)$ were determined from the saturation corrected areas in the Mössbauer spectra. The temperature dependence of $-\ln f(^1A_1)$ follows closely the Debye model with $\Theta_D(^1A_1) = 165$ K; the same holds for $-\ln f(^5T_2)$ only above ~ 210 K with $\Theta_D(^5T_2) = 134$ K. The difference in the Debye-Waller factors implies the lattice being more rigid in the $^1A_1$ phase than in the $^5T_2$ phase.

Whereas the drastic difference in magnetism between the unsolvated $[Fe(papt)_2]$ complex on the one hand, and its solvates with C_6H_6 and 4/3 $CHCl_3$ on the other, are believed to be due to differences introduced into the lattice characteristics by incorporation of the solvent molecules, it could be demonstrated by Keller et al. [157] that the spin transition in $[Fe(papt)_2]$ must be considered largely an intramolecular effect. In the temperature range, where the solution ^{1}H-NMR spectra as well as the solid state magnetic susceptibilities exhibit gross anomalies, the IR spectrum did not change. Thus a structural phase transition accompanying the spin transition should be ruled out here. The lattice energetics, however, appears to govern the fraction α of permanently $^5T_2$ molecules. The lattice free energy reaches a minimum at fairly large values of α in some compounds, such as the crystalline solvates of $[Fe(papt)_2]$; in other cases the spin conversion goes to completion to reach a minimum in the lattice free energy. This has been further substantiated by ^{57}Fe Mössbauer effect measurements on quickly frozen solutions of $[Fe(papt)_2]$ in DMF/glycerol, in chloroform, and in methanol, respectively [158]. In all these matrices practically complete conversion to the $^1A_1$ state has been observed below the glass transition point T_g (~ 160 K in case of the DMF/glycerol matrix), contrary to the incomplete spin transition in the crystalline state. The Debye-Waller factor shows a discontinuity at T_g. An increased transformation $^1A_1 \rightarrow {}^5T_2$ above T_g has been attributed to crystal formation. When a DMF/glycerol solution was slowly cooled below the melting point (~ 200 K) evidently crystallization took place and the ratio of the spin state fractions was observed to adjust, at each temperature, approximately to the fraction found in the solid.

Further Mössbauer effect studies (304 − 12 K) and magnetic susceptibility measurements (301 − 1 K) on the neutral complex [Fe(papt)$_2$] have been performed recently [159]. The magnetic data are shown in Fig. 34. The values of $- \ln f(^5T_2)$ and $- \ln f(^1A_1)$ have been found to follow the high-temperature approximation of the Debye model above 105 K and 140 K, respectively, if anharmonic corrections have been introduced. No simple model is available at present, which would be capable to account for the complete temperature dependence of the Debye-Waller factors in this crossover system.

The effect of the intra-ligand substitution on the magnetic behaviour of the neutral complexes [Fe(Y-papt)$_2$] · Sol, where Y = 3−CH$_3$ (unsolvated), 4−CH$_3$ (unsolvated), 4−CH$_3$ (Sol = 2 C$_6$H$_6$), 4−CH$_3$ (Sol = CHCl$_3$), 6−CH$_3$ (Sol = CHCl$_3$), was studied by Goodwin and Mather [152]. These systems all show $^5T_2 \rightleftharpoons {}^1A_1$ transitions, most of them are incomplete with rather high residual HS fractions at the low temperature end. Some representative $\mu(T)$ data are listed in Table 7. The systems [^{57}Fe(3−CH$_3$-papt)$_2$]·$\frac{1}{6}$ CHCl$_3$ and [^{57}Fe(4−CH$_3$-papt)$_2$] · CHCl$_3$ (with enriched ^{57}Fe) have recently been studied by Mössbauer spectroscopy [185] both in the solid state and in frozen solution (DMF/glycerol and methanol/glycerol, respectively). The Mössbauer spectra confirm the temperature induced interconversion between the $^5T_2$ and $^1A_1$ states. The area fraction of the HS state, at a given temperature, is lower in the frozen solution than in the solid state of both systems. $V_{zz}(^5T_2)$ has been found to be positive for both complexes by Mössbauer measurements in applied magnetic fields.

Effects of the replacement of a hydrogen atom by a methyl group in various positions of the pyridylamino ring of the papt H ligand, as well as the influence of the

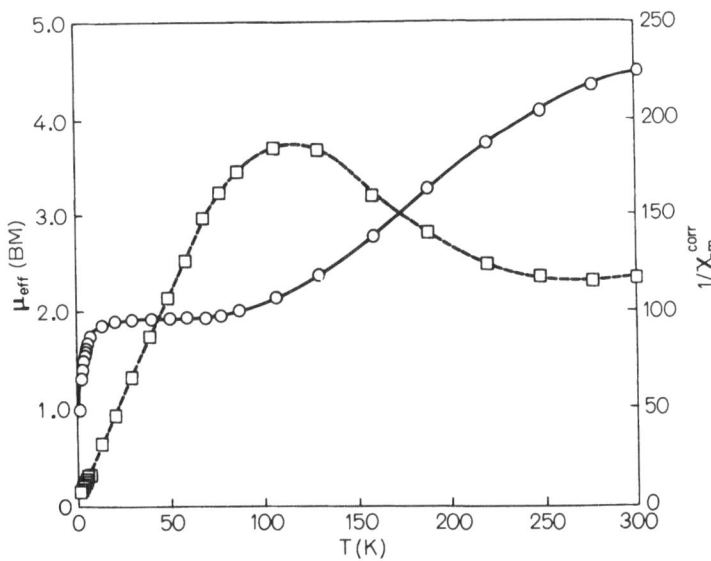

Fig. 34. Temperature dependence of the effective magnetic moment μ_{eff} (○) and the reciprocal molar magnetic susceptibility $1/\mathrm{xm}^{\mathrm{corr}}$ (□) for [Fe(papt)$_2$] (from Ref. 159)

nature of the anion on the magnetism of $[Fe(Y\text{-}papt\,H)_2\,]X_2 \cdot nH_2O$ complexes[152] as compared to the unsubstituted $[Fe(paptH)_2\,]X_2 \cdot nH_2O$ complexes can be seen from the relevant magnetic data of Table 7. In accordance with earlier suggestions[109], the anion dependence is believed to arise mainly from lattice effects, rather than from any intrinsic characteristics of the actual anions. The electronic spectra of the $[Ni(paptH)_2\,]X_2 \cdot nH_2O$ complexes show[152], that the field strengths of the ligands are in the neighbourhood of the crossover region of iron(II).

The influence of the nature of the anion on the $^5T_2(O_h) \rightleftharpoons {}^1A_1(O_h)$ behaviour in $[Fe(paptH)_2\,]X_2 \cdot nH_2O$ which has already been established by magnetic susceptibility measurements[109] (see Table 7), has also been demonstrated by Mössbauer effect studies[160]. The spectra prove the coexistence of the $^5T_2$ and $^1A_1$ states in all instances with $X = Cl^- (2H_2O)$, $Br^- (2H_2O)$, $NO_3^- (H_2O)$, NO_3^-, $ClO_4^- (H_2O)$. The residual $^5T_2$ contribution observed at cryogenic temperatures is nearly zero for the halides and nitrates, but is as high as $\sim 88\%$ in the perchlorate.

The time dependence of the anhydrous $[Fe(paptH)_2\,](NO_3)_2$, which was first reported on by Sylva and Goodwin[109], has recently been studied in detail by ^{57}Fe Mössbauer measurements[161]. The gradual conversion from $^5T_2(O_h)$ to $^1A_1(O_h)$ was found to be almost complete, when the sample was cooled slowly, the $^5T_2$ fraction being > 0.95 at 290 K and 0.067 at 105 K. A pronounced hysteresis of $\Delta T_c = 34$ K has been observed, the transition being centered at $T_c^> = 263$ K for rising tem-

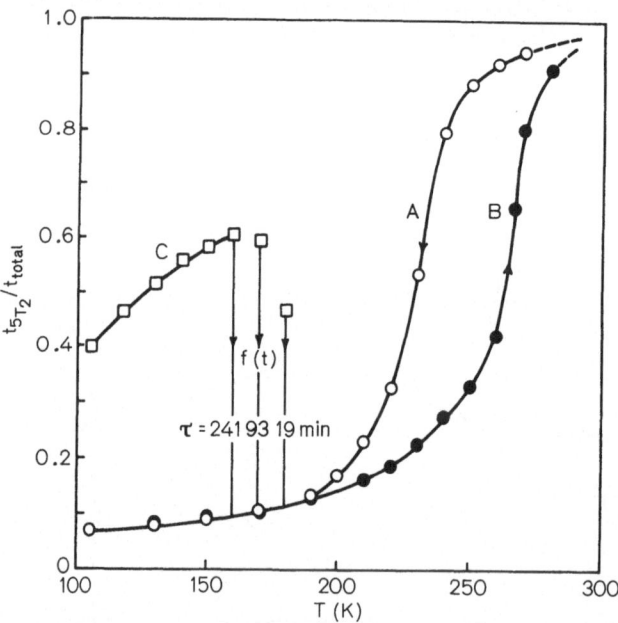

Fig. 35. Temperature dependence of the relative effective thickness $t(^5T_2)/[t(^5T_2) + t(^1A_1)]$ of $[Fe(paptH)_2\,](NO_3)_2$ (A) with slowly decreasing temperature followed by (B) increasing temperature; (C) with increasing temperature after the sample had been rapidly cooled in liquid nitrogen (from Ref. 161)

perature and at $T_c^< = 229$ K for lowering temperature (see Fig. 35). When the sample was cooled rapidly, a considerable fraction (~ 0.40) of the $^5T_2$ state molecules was found to be metastable at 105 K. Subsequent heating caused a $^5T_2(O_h) \rightarrow {}^1A_1(O_h)$ transformation to adjust to the "equilibrium" composition. The kinetics of this transformation were studied at 160, 170, and 180 K; mean lifetimes for the $^5T_2$ state were evaluated to be 14 460, 5 590, and 1 160 s. The activation energy for the transformation was found to be $\Delta E = 7.5$ kcal mol^{-1}. At $T \geqslant 200$ K, the $^5T_2 \rightarrow {}^1A_1$ transformation was always very fast, and consequently, non-equilibrium $^5T_2$ fractions were not observed. In no other spin crossover-system studied have the authors found similar time effects.

Spin state lifetimes in aqueous solutions of [Fe(paptH)$_2$]Cl$_2 \cdot 2$H$_2$O have been measured by Beattie et al.[162] using a resonance and pulsed ultrasonic technique. The relaxation times at 25 °C were found to be on the order of 40 ns.

7.1.3 Other Tri-imine Ligand Complexes with Fe(NN'N)$_2$ Core

Another type of tri-imine complexes containing terpyridyl type ligands, in which the inner pyridyl ring has been replaced by an amino moiety, such as di-(2-pyridylmethyl)-amine *(dpma)* and di-(2-pyridylmethyl)methylamine *(medpma)* has been described

R = H: dpma
R = CH$_3$: medpma

by Nelson and Rodgers[162]. It appears that the crossover point is just inbetween the two ligands, because the [Fe(dpma)$_2$]$^{2+}$ have been found to be low-spin with magnetic moments less than 0.8 B.M., whereas the corresponding iron(II) complexes with *medpma* are high-spin. The strong field nature of dpma is confirmed by the high value (1 255 cm^{-1}) for Dq in [Ni(dpma)$_2$]$^{2+}$ complexes. The authors have interpreted the difference in magnetic behaviour in terms of steric effects and of differences in metal-ligand π-bonding.

Similar terpy-like tri-imine ligands are N'-(2-pyridylmethyl)picolinamidine (ppa) and its two methyl-substituted derivatives meppa and me$_2$ppa[163]. These ligands also form six-coordinate bis-chelate complexes of nickel(II) and iron(II). Magnetic

R^1 = H, R^2 = H: ppa
R^1 = Me, R^2 = H: meppa
R^1 = Me, R^2 = Me: me$_2$ppa

and Mössbauer effect studies show that the ppa complexes of iron(II), [Fe(ppa)$_2$]-X$_2 \cdot n$H$_2$O, with X = ClO$_4^-$ (2H$_2$O) and PF$_6^-$ (H$_2$O), exist in a temperature dependent high-spin ($^5T_2$) $\rightleftharpoons$ low-spin ($^1A_1$) equilibrium, in accordance with predictions based on the ligand field spectra of the nickel(II) complexes. The room temperature magnetic moments of all the [Fe(ppa)$_2$]X$_2 \cdot n$H$_2$O complexes are still close to 1 B.M., indicating that the magnetism is determined by the $^1A_1$ ground state. This is also supported by the Mössbauer parameters. Above room temperature, however, the magnetic moments rise in all cases. But only the perchlorate and hexafluorophosphate could be studied up to $\sim 400\,K$ without apparent change (decomposition and loss of crystal water) of the sample. The PF_6^- salt is virtually high-spin at $\sim 400\,K$. This compound showed some hysteresis in $\mu_{eff}(T)$. Definitely, the magnetic behaviour of the $[Fe(ppa)_2]X_2 \cdot nH_2O$ complexes above room temperature are somewhat dependent on the nature of the anion. The plots of $\ln K$ vs. $1/T$ for the ClO_4^- and the PF_6^- salts are markedly curved and yield relatively high ΔH and ΔS values (for the straight line region between 280 and 330 K), from which it has been concluded that the process involves more than just a simple change in spin. The iron(II) complexes of the substituted ligands meppa and me_2ppa are fully high-spin at all temperatures studied.

7.2 Complexes with Fe(N'NN')₂ Core

Livingstone and Nolan [164] have prepared the tri-imine ligand 2,6-(dibenzothiazol-2-yl)pyridine *(dbtp)*, which differs from 2,2',2''-terpyridyl *(terpy)* in that it has benzo-thiazolyl moieties in place of the outer two pyridyl rings. Like *terpy, dbtp* coordinates

(dbtp)

to a transition metal atom with all three nitrogen atoms in the one plane. Hexa-co-ordinate complexes of the type $[M(dbtp)_2]X_2 \cdot nH_2O$ with $[M(NN'N)_2]$ core as well as five-coordinate species $[M(dbtp)X_2]$ are known. The latter ones are all high-spin. Of interest here is the magnetism of the $[Fe(dbtp)_2]X_2 \cdot nH_2O$ complexes. Benzo-thiazole [165] (basic $pK_a = 1.2$) is a much weaker base than pyridine ($pK_a = 5.2$)[166]. As, in addition, the *dbtp* ligand exerts higher steric requirements causing longer iron-ligand bond lengths than in case of *terpy*, one expects the *dbtp* ligand to form less stable complexes with a weaker ligand field strength than *terpy*. This has in fact been observed [164]: $[Fe(dbtp)_2](ClO_4)_2 \cdot 3H_2O$ (reddish purple at room temperature) shows a gradual $^5T_2(O_h) \rightleftharpoons {}^1A_1(O_h)$ spin transition (some magnetic data are listed in Table 7), whereas $[Fe(terpy)_2]Br_2 \cdot H_2O$ is diamagnetic at room temperature [167]. The solid state UV-vis spectrum of $[Fe(dbtp)_2](ClO_4)_2 \cdot 3H_2O$ shows two weak broad bands at 9300 and 12000 cm^{-1}, which are considered to arise from the $^5E_g \leftarrow {}^5T_2$ transition of high-spin octahedral iron(II), perturbed by low-symmetry field components.

7.3 Five-coordinate Complexes with Fe(PNP)₂ Core and Related Systems

Five-coordinate complexes of the type $[M(pnp)X_2]$ and $[M(pnp)XY]$ (M = Fe(II), Co(II), Ni(II); X = Cl$^-$, Br$^-$, I$^-$; Y = NCS$^-$), differing broadly in their magnetic properties, have been described by Kelly, Ford, and Nelson [168]. *pnp* denotes the

$Ph_2P-CH_2-CH_2$⎯⎯N⎯⎯$CH_2-CH_2-PPh_2$ (pnp)

ligand 2,6-bis-(2-diphenylphosphinoethyl)pyridine. Distorted trigonal bipyramidal structures were assigned on the basis of physical properties. The complexes $[Fe(pnp)Cl_2]$ and $[Fe(pnp)Br_2]$ were found to be fully high-spin, with magnetic moments of 5.3–5.4 B.M. between 293 and 93 K (see Table 7). The magnetic moments of $[Fe(pnp)I_2]$ and $[Fe(pnp)I(NCS)]$, however, vary considerably with temperature (see Table 7), and the authors suggest that a temperature dependent $^5E(D_{3h}) \rightleftharpoons {}^3A(D_{3h})$ transition is involved in these two complexes. Two of the cobalt(II) complexes studied, viz. $[Co(pnp)Br_2]$ and $[Co(pnp)I_2]$, are also proposed to exhibit spin crossover, $^2E \rightleftharpoons {}^4A$, on the basis of their temperature dependent magnetic moments [168]; $[Co(pnp)Cl_2]$ has been found to exist in the ^4A state. It is interesting to note that nickel(II), embraced by the $(pnp)Cl_2$ ligand sphere, undergoes temperature dependent $^1A \rightleftharpoons {}^3E$ transition, as derived from the strong temperature variation of μ_{eff}. $[Ni(pnp)Br_2]$ and $[Ni(pnp)I_2]$, however, are diamagnetic. Thus, the tendency of spin-pairing seems to follow the order Fe < Co < Ni. There is a definite influence of the coordinated halide ions, and it appears that strong covalent interactions, leading to reductions in the interelectronic repulsions, is a major, if not dominant, factor in determining the spin multiplicity. Sacconi [169] has successfully correlated the spin multiplicity of five-coordinate cobalt(II) and nickel(II) complexes with (a) the overall electronegativities and (b) the overall nucleophilic reactivity constants of the five donor atoms.

The tridentate ligand 2,6-bis-(2-diphenylphosphinomethyl)pyridine, *pmp*, forms similar five-coordinate $[M(pmp)X_2]$ complexes as the *pnp* ligand [170]. The molecular

Ph_2P-CH_2⎯N⎯CH_2-PPh_2 (pmp)

structure, however, has been suggested to be square-pyramidal on the basis of their electronic spectra and other physical properties. The iron(II) complexes $[Fe(pmp)X_2]$ ($X = Cl^-$, Br^-, I^-, NCS^-) are all high-spin. Two cobalt(II) complexes have been found to show spin isomerism $^2A_1(C_{4v}) \rightleftharpoons {}^4A_2(C_{4v})$, viz. $[Co(pmp)Br_2]$ and $[Co(pmp)(NCS)_2]$; the iodo complex $[Co(pmp)I_2]$ is diamagnetic. All three nickel(II) halide complexes, $[Ni(pmp)X_2]$ ($X = Cl^-$, Br^-, I^-), were found to be diamagnetic, whereas $[Ni(pmp)(NCS)_2]$ shows $^1A_1(C_{4v}) \rightleftharpoons {}^3B_1(C_{4v})$ crossover. The effect of a change in structure from a distorted trigonal bipyramid to a distorted square pyramid, as reflected by differences in the X-ray pattern, on the position of the magnetic crossover seems to be of special interest in these complexes. The tendency of spin-pairing follows again the order Fe < Co < Ni. As with the *pnp* complexes, spin-pairing is promoted by an increase of the polarizability of the halide ion; the stability of the complexes also increases in the order Cl < Br < I.

Sacconi and his school have prepared a large number of five-coordinate complexes containing a tetradentate "tripod" ligand with various donor sets. They have found that the spin multiplicity of five-coordinate iron(II) complexes decreases from five to one with increasing overall nucleophilicity and with decreasing overall electronegativity of the donor atoms[171]. Depending on the nature of the donor set, one can distinguish between four classes of iron(II) complexes with different spin states:

(i) S = 2 complexes with $N_4X(X = halide)$, NP_2X_2, and NP_3X donor sets;
(ii) complexes with N_2P_2I and NP_2I_2 donor sets showing $S = 2 \rightleftharpoons S = 1$ crossover;
(iii) S = 1 complexes with P_4X donor sets;
(iv) complexes with P_4X sets showing $S = 1 \rightleftharpoons S = 0$ crossover.

8 Other Crossover Systems of Iron(II)

8.1 Bis(hydrotris(pyrazolyl)borato)iron(II)

Already in the early stage of spin crossover research on iron(II) complexes, Jesson, Weiher, and Trofimenko[172,173] have found that octahedral ferrous complexes based on the hydrotris(1-pyrazolyl)borate, [Fe(HB)X,Y-pz)$_3$)$_2$] (see Fig. 36) can show considerable differences in magnetism produced by minor changes in ligand substitution. Similar behaviour occurs in solution[174]. In the crystalline state, [Fe(HB(me-pz)$_3$)$_2$] (I) shows temperature dependent $^5T_2(O_h) \rightleftharpoons {}^1A_1(O_h)$ crossover, whereas [Fe(HB(pz)$_3$)$_2$] (II) is fully low-spin, and the dimethyl analog [Fe(HB(me$_2$-pz)$_3$)$_2$] (III) is high-spin at all temperatures studied[173]. The variable-temperature Mössbauer spectra of compound I reflect the coexistence of the two spin states, $^5T_2(O_h)$ and $^1A_1(O_h)$, the intensity of the HS state decreasing with decreasing temperature at the favour of the LS state (see Fig. 37). From a ligand field calculation for the quintet states, including spin- orbit coupling and trigonal distortion, the orbital splitting of

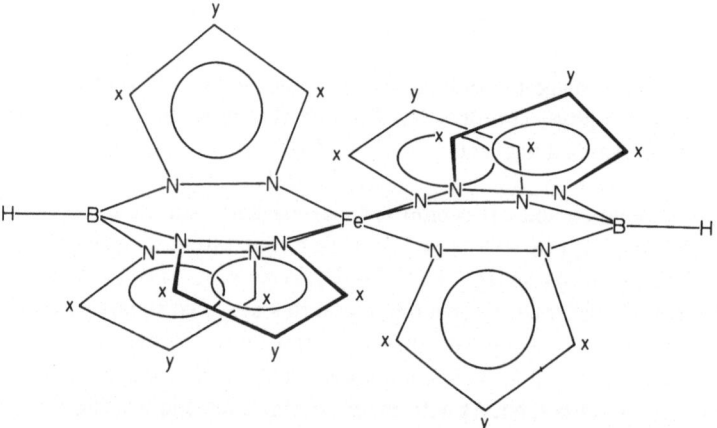

Fig. 36. Chelate structure of octahedral iron(II) complexes based on the hydrotris(1-pyrazolyl)-borate, HB (X, Y-pz)$_3$, ligand. (I) X = CH$_3$, Y = H: [Fe(HB(me-pz)$_3$)$_2$]; (II) X = Y = H: [Fe(HB(pz)$_3$)$_2$]; (III) X = Y = CH$_3$: [Fe(HB(me$_2$-pz)$_3$)$_2$] (from Ref. 172)

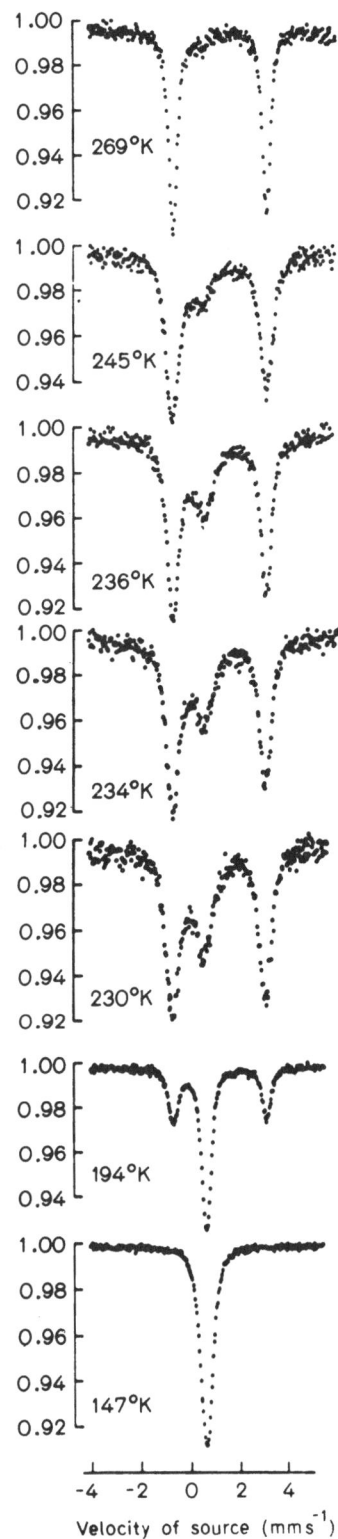

Fig. 37. ^{57}Fe Mössbauer spectra of [Fe(HB(me-pz)$_3$)$_2$] at various temperatures. The outer two lines refer to the $^5T_2(O_h)$ doublet. The inner line is an unresolved quadrupole doublet arising from the $^1A_1(O_h)$ state (from Ref. 173)

the high-spin ground state due to the trigonal distortion is $\sim 1000\ cm^{-1}$ leaving the orbital singlet lowest. The authors used the Van Vleck formalism to account for the anomalous temperature dependence of the magnetic susceptibility of compound I. Agreement with experimental data could not be reached by using a fixed energy level separation between the $^1A_1$ state and the center of gravity of the $^5T_2(O_h)$ manifold; the authors ascribe this failure to a possible change in the crystal structure.

In solution, compound II exhibits a smooth temperature-dependent $^5T_2(O_h) \rightleftharpoons {}^1A_1(O_h)$ transition, whereas compounds I and III behave as normal paramagnetic complexes [174]. Spin state lifetime measurements on these and other iron(II) spin crossover systems have been carried out by Sutin et al. [175] and Beattie et al. [176]; the results are given in Chap. 9.

8.2 Complexes with Dithiazoline and Dihydrothiazine Ligands

A series of iron(II) complexes of the type $[FeL_2X_2]$ has been described [177], where L denotes the bidentate α-diimine ligand 2,2'-bi-4,5-dihydrothiazine (btz) or 2,2'-bi-2-thiazoline (bt) or its alkyl-substituted derivatives (bts, btn, and btne); X refers to

bt: R=R'=H
bts: R=H, R'=Me
btn: R=Me, R'=H
btne: R=Et, R'=H

NCS^- or $NCSe^-$. The complexes were characterized, by IR and electronic spectra together with electrical conductance measurements, as having neutral (probably cis) six-coordinate structures, in which the organic ligands as well as the pseudohalide anions are coordinated via the nitrogen atoms. Four of the seven complexes under study [177] show temperature dependent changes of the spin states, $^5T_2(O_h) \rightleftharpoons {}^1A_1(O_h)$, with extremely sharp transitions in case of $[Fe(bt)_2(NCS)_2]$ $(T_c \sim 180\ K)$ and $[Fe(bt)_2(NCSe)_2]$ $(T_c \sim 200\ K)$, and somewhat less abrupt transitions in case of $[Fe(bts)_2(NCS)_2]$ $(T_c \sim 220\ K)$ and $[Fe(btz)_2(NCS)_2]$ $(T_c \sim 220\ K)$ (see Fig. 38). Some characteristic magnetic data are listed in Table 8. Hysteresis effects are also reported in Ref. [177]. There is a striking change of the spin state behaviour on going from NCS^- to $NCSe^-$ in the btz complex, which has very rarely been seen. Most probably, the ligand field strength in the $[Fe(btz)(NCSe)_2]$ complex is already very close to the crossover point, and only a slight reduction of the basicity of the nitrogen atom on going to the NCS^- ligand suffices to induce the thermal population of both spin states. The 93 K data of μ_{eff} of the two complexes also indicate that the average ligand field seems to be slightly smaller in $[Fe(btz)_2(NCS)_2]$. This difference is well substantiated by the electronic spectra of the analogous nickel(II) complexes: the $^3A_{2g} \rightarrow {}^3T_{2g}$ transition in $[Ni(btz)_2(NCS)_2]$ appears at $10\,500\ cm^{-1}$, that in $[Ni(btz)_2(NCSe)_2]$ at $10\,800\ cm^{-1}$ [177].

The complexes of the btn and btne ligands which bear alkyl substituents adjacent to the donor atoms are high-spin at all temperatures studied. The electronic spectra

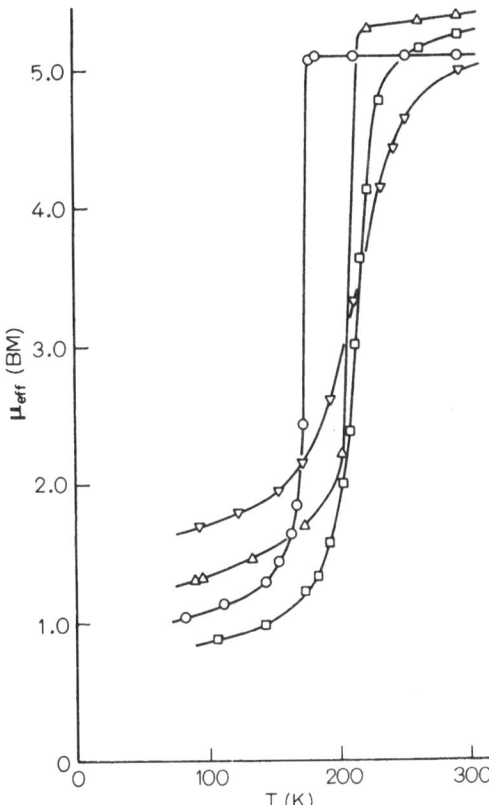

Fig. 38. Temperature dependence of the effective magnetic moment μ_{eff} of $[Fe(bt)_2(NCSe)_2]$ (△), $[Fe(bts)_2(NCS)_2]$ (□), $[Fe(bt)_2(NCS)_2]$ (○), $[Fe(btz)_2(NCS)_2]$ (▽). The data refer to cooling curves (from Ref. 177)

of the corresponding nickel(II) complexes again indicate a correlation between the apparent Dq(Ni) values and the ground state of the iron(II) complexes. However, the critical crossover value for Dq(Ni) of $\sim 10\,500$ cm^{-1} is much lower for this series of neutral mixed-ligand complexes than previously found for cationic octahedral bis-(tridentate ligand) complexes[162] and for cationic tris(bidentate ligand) complexes[178].

Table 8. Magnetic data of iron (II) complexes with dithiazoline and dihydrothiazine ligands

Compound	T(K)	μ_{eff} (B.M.)	Ground state[a]	Ref.
$[Fe(btz)_2(NCS)_2]$	293/93	5.01/1.68	$^5T_2 \rightleftharpoons {}^1A_1$	177
$[Fe(btz)_2(NCSe)_2]$	293/93	0.50/0.45	$^1A_1$	177
$[Fe(bt)_2(NCS)_2]$	293/93	5.10/1.05	$^5T_2 \rightleftharpoons {}^1A_1$	177
$[Fe(bt)_2(NCSe)_2]$	293/93	5.40/1.31	$^5T_2 \rightleftharpoons {}^1A_1$	177
$[Fe(bts)_2(NCS)_2]$	293/93	5.26/0.84	$^5T_2 \rightleftharpoons {}^1A_1$	177
$[Fe(btn)_2(NCS)_2]$	293/93	5.27/5.21	$^5T_2$	177
$[Fe(btne)_2(NCS)_2]$	293/93	5.27/5.13	$^5T_2$	177

[a] Term symbols given in the approximation of O_h symmetry

The very abrupt spin transition in [Fe(bt)$_2$(NCS)$_2$] has extensively been studied by König et al.[124,184] using Mössbauer spectroscopy and X-ray powder diffraction. A pronounced hysteresis of $\Delta T_c = 9.5$ K has been observed ($T_c = 181.86$ K for rising and $T_c = 172.33$ K for falling temperature). The authors attempted to describe the results with the thermodynamic model for spin transitions of Slichter and Drickamer [97]. Different diffraction patterns were recorded above and below T_c, where only one spin isomer was present (see Fig. 39). This provides evidence that a crystallographic phase change accompanies the abrupt spin phase transition in this compound. The temperature dependence of the peak profiles was found to follow that of the HS fraction as derived from the Mössbauer spectra, which shows that the crystallographic phase change is directly associated with the interconversion of the two spin phases. The Debye-Waller factors were evaluated for the two spin phases; they showed at T_c a discontinuity of $\Delta f \sim 35\%$ on going from the HS to the LS phase. The shape of the transition curve near T_c and T_c itself, were somewhat different in three independently prepared samples.

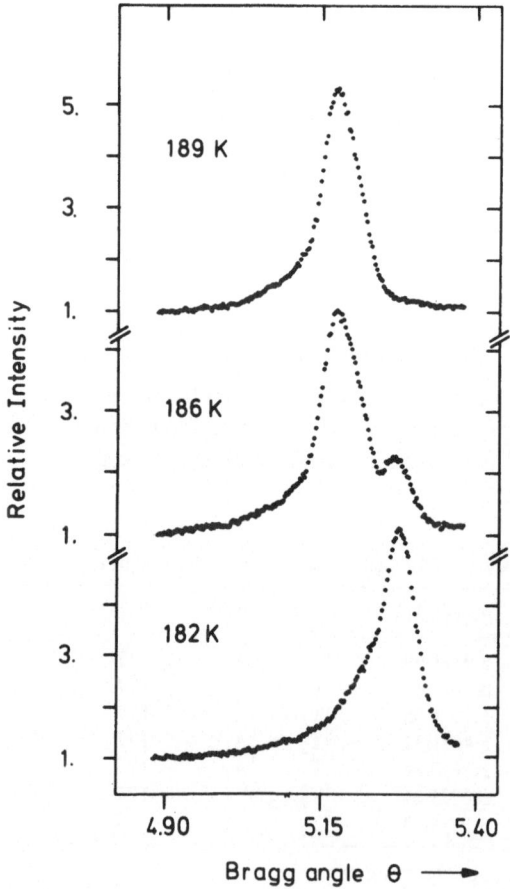

Fig. 39. Peak profiles of X-ray powder diffraction on [Fe(bt)$_2$-(NCS)$_2$] in the temperature region of the spin transition (from Ref. 124)

8.3 Phosphine Complexes

Nelson and coworkers [187] have synthesized a series of mixed-ligand iron(II) complexes of the type $[Fe(dippen)_2 X_2]$, where *dippen* is cis-1,2-bis(diphenylphosphino)-ethylene, and $X = Cl^-, Br^-, NCS^-$, and N_3^-. Temperature dependent magnetic sus-

H   H
\   /
 C=C
/   \
Ph_2P PPh_2 dippen

ceptibility (ca. 350 – ca. 80 K) and Mössbauer effect measurements (298 – 77 K) have classified the complexes as having LS ($X = NCS^-, N_3^-$) or HS ($X = Br^-$) ground states. $[Fe(dippen)_2 Cl_2]$ has been found to exhibit HS $\rightleftharpoons$ LS transition, with magnetic moments ranging from 4.84 B.M. at 353 K to 3.41 B.M. at 83 K. The Mössbauer spectra confirm the coexistence of the two spin states, S = 2 and S = 0, respectively.

The partial ligand field strengths of phosphine ligands are known to be somewhat larger than those of nitrogen donor ligands. This explains the LS behaviour of the present $[Fe(dippen)_2 X_2]$ complexes with $X = NCS^-, N_3^-$, whilst many mixed-ligand complexes of the type $[Fe(diimine)_2 (NCS)_2]$ show temperature dependent HS $\rightleftharpoons$ LS transition. The observed spin crossover in the present $[Fe(dippen)_2 Cl_2]$ complex is clearly due to the weaker partial ligand field strength of Cl^- as compared to NCS^- and N_3^-, which reduces the strong-field property of *dippen* such as to shift the overall ligand field strength into the crossover region.

$[Fe(dippen)_2 Cl_2]$ is the first, and to our knowledge so far only, example of spin crossover in iron(II) phosphine complexes.

8.4 Complexes with Hexadentate Ligands

In 1975 Wilson et al. [37] have communicated the synthesis and spectral characterization of a series of pseudooctahedral iron(II) complexes of the hexadentate ligand tris[4-[(6-R)-2-pyridyl]-3-aza-3-butenyl]amine, where R is either H or CH_3, abbreviated as $[Fe(6Mepy)_n(py)_m tren]$ where *tren* is $N(CH_2 CH_2-NH_2)_3$. The magnetism of the following four complexes has been studied in the solid state as well as in solution (acetone and DMSO, respectively):

I: $[Fe(py)_3 tren](PF_6)_2$; R = R' = R'' = H

II: $[Fe(6Mepy)(py)_2 tren](PF_6)_2$; R = R' = H, R'' = CH_3

III: $[Fe(6Mepy)_2(py)tren](PF_6)_2$; R = H, R' = R'' = CH_3

IV: $[Fe(6Mepy)_3 tren](PF_6)_2$; R = R' = R'' = CH_3

where the cationic complex has the structure.

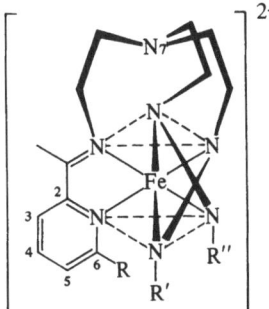

In the solid state, II, III, and IV show $^1A_1 \rightleftharpoons {}^5T_2$ spin equilibrium, while I is low-spin. In solution, II and III retain the spin equilibrium property, whereas IV is fully high-spin and I is low-spin over a temperature range of approximately 200 degrees. From the electronic spectra the critical field strength (crossover point) has been estimated to lie near $11\,700$ cm^{-1}. No solvent dependence has been observed.

The effect of intraligand substitution on the spin crossover behaviour in these systems has been further investigated using various techniques [188]. Variable temperature Mössbauer spectra show the quadrupole doublets of the two coexisting spin states. Thermodynamic parameters have been derived from the temperature dependent magnetic susceptibility data. ΔH has been found to be 4.8, 3.1, and 4.8 kcal · mol^{-1} for II, III, and IV, serially, in the solid state, and 4.6 and 2.8 kcal mol^{-1} for II and III, respectively, in solution.

The evaluated entropy changes are $\Delta S = 9.7, 9.7$, and 20.9 eu for II, III, and IV, serially, in the solid state, and 10.0 and 8.5 eu for II and III, respectively, for the spin transition in solution.

An X-ray structural study of IV performed by Delker and Stucky (private communication in Ref. 188) indicates that a steric interaction between the methyl groups and adjacent pyridine rings is largely responsible for the observed changes in the magnetic behaviour of the series I, II, III, and IV. Furthermore, X-ray studies at variable temperature for IV reveal an average decrease of ca. 0.12 Å in the six Fe-N bond distances in going from room temperature ($\mu \sim 5.0$ B.M.) to 205 K ($\mu \sim 2.3$ B.M.). The authors have pointed out that, particularly with multidentate ligands, it should be possible to "fine-tune" the ligand field strength by step-wise substitution of hydrogen by methyl groups to produce gradual perturbations through both steric and electronic interactions.

The frequently observed influence of the non-coordinated anion on the spin crossover characteristics has also been encountered in these hexadentate systems [188].

The spin state lifetimes in solution of the complexes II and III have been measured directly with the laser Raman temperature-jump technique [189]. Changes in the absorbance at ~ 560 nm (CT band maximum) following the T-jump perturbation indicate that the relaxation back to equilibrium occurs by a first-order process. The spin-state lifetimes are $\tau(\text{LS}) = 2.5 \cdot 10^{-6}$ s and $\tau(\text{HS}) = 1.3 \cdot 10^{-7}$ s. The enthalpy change is $\Delta H \leqslant 5$ kcal mol^{-1}, in good agreement with that derived from $\chi(\text{T})$ data in Ref. 188. The dynamics of intersystem crossing processes in solution for these hexadentate complexes and other six-coordinate d^5, d^6, and d^7 spin-equilibrium complexes of iron(III), iron(II), and cobalt(II) has been discussed by Sutin and Wilson et al. [110].

ESCA studies on these hexadentate complexes by Lazarus, Hoselton, and Chou [191] were not successful. It has been found that the broad satellite structure in the observed 2 p X-ray photoelectron spectra were due to a radiation-induced decomposition of the HS isomer in the spectrometer rather than the result of the multiplet splitting.

8.5 Complexes with Tetradentate Macrocyclic Ligands

Busch and coworkers[192,193] have reported on the synthesis, magnetism, IR, and
electronic spectral properties of iron(II) complexes containing tetradentate macrocyclic nitrogen donor ligands.

The iron complexes of the ligand 5,7,7,12,14,14-hexamethyl-1,4,8,11-tetra-
azacyclotetradeca-4,11-diene, abbreviated as [14]dieneN$_4$, generally fall into three
categories:

(i) five-coordinate HS iron(II);
(ii) six-coordinate LS iron(II);
(iii) six-coordinate LS iron(III).

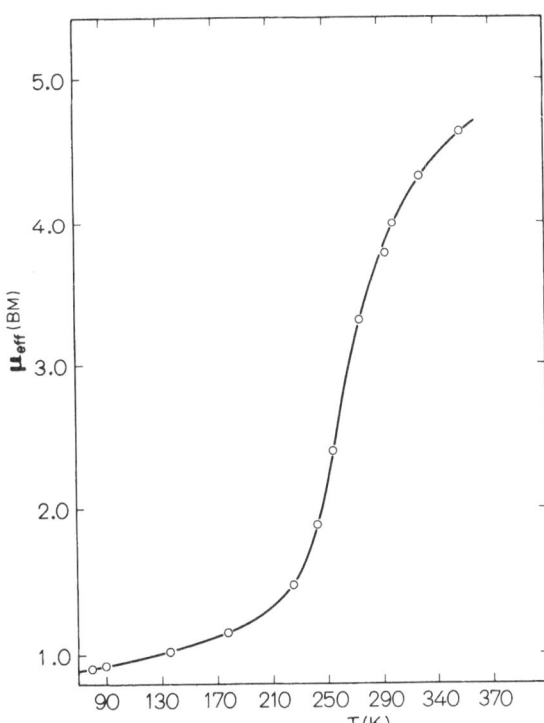

[14]dieneN$_4$

One complex, a phenanthroline adduct, [Fe([14]dieneN$_4$)phen](ClO$_4$)$_2$ does not fit
into this scheme in that it shows S = 2 $\rightleftharpoons$ S = 0 spin equilibrium at room temperature.
The magnetic moment as a function of temperature between ~ 80 K and ~ 350 K is
reproduced in Fig. 40. A plot of log K vs. 1/T, where the equilibrium constant is calculated in terms of the relative concentrations of the HS and the LS isomers, as

Fig. 40. Magnetic moment as
a function of temperature for
the macrocyclic ligand complex
[Fe([14]dieneN$_4$)phen](ClO$_4$)$_2$
(from Ref. 192)

determined from the magnetic susceptibilities as a function of temperature, yields a straight line over the range 350 – 180 K. For this temperature range the enthalpy change ΔH and the entropy change on going from LS to HS has been evaluated to be 6.58 kcal mol^{-1} and 22 eu, respectively. These values are considerably larger than those expected for a change in spin state alone. The changes in bond lengths and angles accompanying the change in spin state, as has been confirmed by single-crystal X-ray analysis on a number of spin crossover systems [30,31,37,76,180], well accounts for the relatively large ΔH and ΔS values.

Five-coordinate HS iron(II) complexes of the type [Fe([14]dieneN$_4$)X]ClO$_4$ are formed if weak axial ligands (X = Cl, Br, I) are present. Six-coordinate LS iron(II) complexes are formed when ligands of moderate to strong ligand-field strength (X = NCS, acetonitrile, imidazole) are available for coordination in axial sites. The change in the coordination number from five in case of HS iron(II) complexes to six in case of LS iron(II) and iron(III) complexes is well explained by the decrease of the size of the metal ion in going from HS iron(II) to LS iron(II) and LS iron(III). The smaller the central ion the more easily will it be accomodated in the macro-cyclic ligand plane allowing six-coordination to be formed; the larger ion is pushed out of the ring and a 5-coordinate square-pyramidal structure will be favoured.

[14]aneN$_4$

Busch and coworkers [193] have also synthesized and characterized six-coordinate iron(II) complexes [Fe([14]aneN$_4$)X$_2$], where [14]aneN$_4$ refers to the 14-membered fully saturated macrocyclic tetradentate ligand meso-5,5,7,12,12,14-hexamethyl-1,4,-8,11-tetraazacyclotetradecane. The complexes are found to be predominantly low-spin, when the axial ligands X are relatively strong; examples, in spectrochemical order, are CN$^-$ > NO$_2^-$ > NCS$^-$ > CH$_3$CN.

These compounds are the first examples for the stabilization of the LS state of iron(II) with saturated nitrogen as the most abundant donors; this has been ascribed by the authors to the "enhanced ligand field strength due to the constrictive effect of the mechanically confining in-plane macrocyclic ligand". Six-coordinate iron(II) complexes are formed with relatively weak axial ligands such as X = CH$_3$COO$_2^-$ > Cl$^-$ > Br$^-$ > I$^-$ > BF$_4^-$. The complex with X = NCS$^-$ (and possibly with NO$_2^-$ and CH$_3$CN) shows spin crossover at elevated temperatures. ΔH and ΔS were estimated to be ~ 7.0 kcal mol^{-1} and 19 eu, respectively, for the NCS$^-$ derivative.

Proton magnetic resonance studies on [Fe([14]aneN$_4$)(NCS)$_2$] in CHCl$_3$/CDCl$_3$ solution as a function of temperature has shown that a $^1A_1(O_h)$ ⇌ $^5T_2(O_h)$ spin equilibrium is operative also in solution [193]. The interconversion between the two spin isomers must be rapid (> 10^5 s^{-1}), since on the PMR time scale only the "average" resonances are observed.

8.6 [FeL₆] Complexes with Monodentate Ligands

Only one iron(II) complex with six monofunctional ligands has become known so far which shows temperature dependent HS $\rightleftharpoons$ LS transition: $[Fe(isoxazole)_6]X_2$, where $X = BF_4^-$ and ClO_4^-. The magnetic moment changes from 1.3 B.M. at 90 K, to 5.28 B.M. at 298 K, with a very sharp transition at 212 K (see Fig. 41). The typical HS and LS resonances of iron(II) appear in the Mössbauer spectra with strongly temperature dependent intensities; additional resonance lines are due to decomposition products and show that the hexakis-isoxazole complex is not very stable. IR and UV-vis spectra suggest that the iron(II) central ion is surrounded by a regular octahedral array of the monofunctional isoxazole ligands.

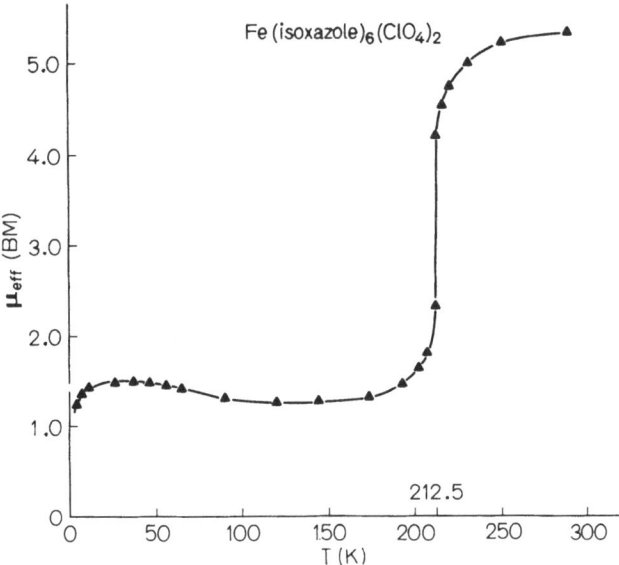

Fig. 41. Magnetic moment as a function of temperature of $[Fe(isoxazole)_6](ClO_4)_2$ (from Ref. 194)

8.7 Inorganic Spin Crossover Systems of Iron(II)

Eibschütz et al.[195-197] have investigated the substitutionally iron-doped layer chalcogenide material 1 T-$Fe_xTa_{1-x}S_2$ with iron concentrations $x \leqslant 1/3$ by means of magnetic susceptibility and Mössbauer effect measurements. They have found that the iron is accomodated exclusively as Fe^{2+} in this system and undergoes a continuous $^1A_1(O_h) \rightleftharpoons {}^5T_2(O_h)$ transition between ~ 200 and ~ 500 K. The transition from the LS to the HS state moves to higher temperatures with increasing x. The fluctuation rate between the two spin states must be faster than 10^7 s^{-1}, because the Mössbauer spectrum is time-averaged at all temperatures studied and only one set of resonance lines is seen. A dynamic transition of this type has never been reported for a metalorganic spin crossover system of iron(II), whereas it is known for metalorganic iron(III) spin crossover systems [198].

The authors have interpreted their data using a statistical model, which reveals that the energy gap between the $^1A_1$ ground state and the lowest of the spin-orbitally split $^5T_2$ levels is temperature dependent; the form of this temperature dependence changes with x. The model includes spin-orbit coupling, low-symmetry field distortion, covalency, and dynamic as well as static effects of local ligand vibrations.

8.8 Singlet ⇌ Triplet Spin Transition in Iron(II) Complexes

Singlet ⇌ Triplet spin crossover in iron(II) complexes has very rarely been observed so far. König, Ritter, Goodwin and Smith [201] have investigated the magnetic behaviour of the solid complex $FeLCl_2 \cdot H_2O$, where L = 1,10-phenanthroline-2-carbothiamide, a tridentate ligand, which coordinates to the iron atom by the two nitrogen atoms of *phen* and the sulphur atom. The water molecule is not coordinated to the iron, but is involved in hydrogen bridging with the free NH_2 group, as concluded from IR spectra [201].

The Mössbauer spectra (between 4.2 K and 298 K) show two overlapping doublets with temperature dependent relative intensities. In conjunction with the data from magnetic susceptibility measurements (between 83 and 313 K) the Mössbauer spectra were interpreted in terms of a singlet ⇌ triplet spin transition of the central iron(II) ion. No internal magnetic field was observed in a Mössbauer experiment under applied field at 4.2 K. The authors consider this complex the first authentic example of a singlet ⇌ triplet spin transition in the 3 d^6 configuration of iron(II). Support has been received from UV-vis spectra and IR vibrational spectra [201].

Other examples of a presumed singlet ⇌ triplet spin transition are the five-coordinated complexes $[Fe(P_4)X]BPh_4$, where (P_4) denotes the tetradentate ligand 1,4,7,10-tetraphosphadecane and X = Br^-, I^-. Bacci, Sacconi et al. [202] have first reported on the unusual magnetic behaviour which they interpreted in terms of a thermally induced singlet ⇌ triplet transition. König, Ritter, and Goodwin have studied these complexes by Mössbauer spectroscopy (between 4.2 and 298 K) [203] and found that the results are consistent with singlet ⇌ triplet transition with a fluctuation rate of $> 10^9 s^{-1}$.

9 Spin Crossover of Iron(II) Complexes in Solution

Some of the polycrystalline spin crossover systems of iron(II) described above retain their spin equilibrium property upon dissolution in appropriate solvents. The Evans NMR method of measuring the change of the paramagnetic shift with temperature is the most common technique to study the magnetic behaviour of such systems. The spin transition characteristics has been observed to depend on various chemical modifications, such as changing a substituent in a ligand or replacing a whole ligand. It is conceivable that these operations will change a number of intramolecular properties (electronic structure, steric hindrance, vibrational modes) and intermolecular properties (solute-solvent interaction, hydrogen bonding, intermolecular vibrational modes), which in turn will have more or less significant bearings on the spin transition characteristics of the dissolved complex. It is believed that intramolecular changes are far more effective than intermolecular ones. This is quite often observed with complexes containing bulky organic ligands, which strongly reduce the influence of the solvent on the electronic structure of the metal ion. Quite often one sees very little or no solvent effects at all. An example is the recently communicated observation of $^5T_2(O_h) \rightleftharpoons {}^1A_1(O_h)$ spin crossover in a cobalt(III) (d^6) complex compound in various solvents [204]; this compound was synthesized and characterized by Kläui as being the first case of quintet $\rightleftharpoons$ singlet spin crossover in a cobalt(III) complex [205,206].

$^5T_2(O_h) \rightleftharpoons {}^1A_1(O_h)$ spin equilibrium in iron(II) complexes based on the hydrotris(1-pyrazolyl)borate ligand, was established for $[Fe(HB(me-pz)_3)_2]$ (for abbreviations see Sect. 8.1) in the solid state [172,173] and for $[Fe(HB(pz)_3)_2]$ in solution [174]. Sutin et al. [175] studied the dynamics of the spin interconversion in CH_2Cl_2/CH_3OH solutions with the laser Raman temperature-jump technique between 0 and 25 °C. The relaxation was observed to be first order with a lifetime of 32 ± 10 ns, independent of temperature and concentration over the range studied. The k_1 and k_{-1} values for the process

$$^1A \underset{k_{-1}}{\overset{k_1}{\rightleftharpoons}} {}^5A \qquad\qquad (9.1)$$

calculated from the measured relaxation time and equilibrium constant are $1 \cdot 10^7$ and $2 \cdot 10^7$ s^{-1}, respectively, at 25 °C. The activation parameters derived from the small temperature dependence of the rate constants are $\Delta H_1^{\ddagger} \leqslant 4$ kcal mol^{-1}, $\Delta H_{-1}^{\ddagger} \leqslant 0$, and $\Delta S_1^{\ddagger} = 8.2$ eu.

Beattie et al. [176] have used resonance and pulsed ultrasonic techniques to determine the relaxation times for the spin equilibrium of $[Fe(HB(pz)_3)_2]$ in tetrahydrofuran solutions. They found a relaxation time of 33 ns at 25 °C independent of concentration. The relaxation amplitudes, however, did depend linearly on concentration, as expected for relaxations of unimolecular processes. The equilibrium constant as a function of temperature was determined from $\chi(T)$ data measured by the Evans' NMR method over a range of -50 to 50 °C. The enthalpy difference $\Delta H^{\circ} =$ H(HS)-H(LS) between the two spin isomers was found to be 5.03 kcal mol^{-1}, compared with 3.85 kcal mol^{-1} found in CH_2Cl_2 solution [174], and the entropy difference $\Delta S = $ S(HS)-S(LS) was calculated to be 13.6 eu, compared with 11.4 eu in CH_2Cl_2 solution [174]. The activation enthalpies $\Delta H_1^{\ddagger}$ and $\Delta H_{-1}^{\ddagger}$ were evaluated to be 5.64 and 0.62 kcal mol^{-1}, respectively, in good agreement with the earlier results of Sutin et al. [175]. An average of 22 cm^3 mol^{-1} was estimated for the volume difference between the LS and HS isomers of $[Fe(HB(pz)_3)_2]$, which probably corresponds to a 0.10–0.15 Å change in the metal-ligand bond length. Using absolute rate theory the authors have estimated a minimum value of $10^{-4.2}$ for the transmission coefficient κ, i.e. the minimum probability for intersystem crossing.

By the same technique of resonance and pulsed ultrasonic perturbation Beattie et al.[176] have also studied the spin crossover dynamics in aqueous solutions of $[Fe(paptH)_2]Cl_2$. Relaxation times of ca. 40 ns were measured at 25 °C, again independent of concentration and thus presumably arising from a unimolecular spin isomerization process. Some of the thermodynamic parameters of particular interest are here: $\Delta H^\circ = 3.91$ kcal mol^{-1}, $\Delta S^\circ = 14.8$ eu, $\Delta H_1^\ddagger = 7.63$ kcal mol^{-1}, $\Delta H_{-1}^\ddagger = 3.72$ kcal mol^{-1}, $\kappa_{min} = 10^{-2.5}$.

Wilson et al.[37] have first reported on the $^5T_2(O_h) \rightleftharpoons {}^1A_1(O_h)$ spin transition taking place in six-coordinate iron(II) complexes with the hexadentate ligand tris-[4-[(6-R)-2-pyridyl]-3-aza-3-butenyl]amine, where R is either H or CH_3 (see Sect. 8.4). Three of these systems, viz. II, III, and IV of Sect. 8.4, show spin crossover in the solid state, and two of them, viz. II and III, also in solution, whereas IV remains fully high-spin and I fully low-spin. The pronounced differences in the magnetic behaviour between the four members of the $[Fe(6-Mepy)_n(py)_m tren](PF_6)_2$ series (see Sect. 8.4) is certainly a consequence of the modification of the ligand by stepwise substituting H for CH_3 and thereby changing essentially the steric hindrance. By comparing the electronic solution spectra of the iron complexes with the analogous nickel(II) complexes the authors arrived at estimates for the cubic ligand field strength and the mean spin pairing energy $\bar{P}$ for the iron(II) complexes. The thermodynamic parameters ΔH° and ΔS° for the spin crossover systems II and III in solution were derived from $\ln K$ vs. $1/T$ plots using the $\chi(T)$ data from Evans NMR method[188]. ΔH° was found to be 4.59 and 2.84 kcal mol^{-1} for II and III, respectively, and ΔS° to be 9.98 and 8.50 eu for II and III, respectively. These values are rather similar to those obtained for $[Fe(HB(pz)_3)_2]$ in solution[174,175,176].

A direct measurement of the spin state lifetimes in hydroxylic solvents (methanol, water, and of mixtures of these with acetone) of the systems II and III of $[Fe(6-Mepy)_n(py)_m tren](PF_6)_2$ has also been performed using the laser Raman T-jump technique[189]. The lifetimes of the ^1A state and the ^5A state are found to be $2.5 \cdot 10^{-6}$ s and $1.3 \cdot 10^{-7}$ s, respectively, in case of II, and $2.5 \cdot 10^7$ s and $2 \cdot 10^{-7}$ s, respectively, in case of III, and thus are significantly longer than for the [Fe(pyrazolylborate)$_2$] complex $(\tau(^1A_1) = 1 \cdot 10^{-7}$ s, $\tau(^5A) = 5 \cdot 10^{-8}$ s)[175]. The authors believe that this might be due to specific non-bonding steric interactions within the hexadentate ligand, which may be energetically restrictive to the primary coordination sphere reorganizational processes known to accompany spin conversion.

The complexes $[Fe(pyim)_3]^{2+}$ and $[Fe(pyben)_3]^{2+}$ were also observed to exhibit the phenomenon of spin crossover in solution ($CH_3CN(20\%)/CH_3OH$ and acetone) by Wilson et al.[145], contrary to an earlier report[142,143]. The molar changes in enthalpy and entropy, ΔH° and ΔS°, were calculated using $\chi(T)$ data from NMR measurements over a temperature of ca. 70–80 K. The results of ΔH° range from 3.7 to 5.1 kcal mol^{-1} and are similar to those found in $[Fe(HB(pz)_3)_2]$ and $[Fe(6-Mepy)(py)_2 tren]^{2+}$. The authors attribute the enthalpic change (of 2–6 kcal mol^{-1}) predominantly to inner-sphere reorganization in the present case as well as in other spin crossover systems in solution. The entropy changes ΔS° of ca. 12 eu for the $[Fe(pyim)_3]^{2+}$ complex and 19–22 eu for $[Fe(pyben)_3]^{2+}$ are much in excess of the electronic contribution $\Delta S_{el} = R \cdot \Delta \ln(2 S + 1)$, which is only 3.2 eu for a $^1A \rightleftharpoons {}^5A$ conversion. The authors ascribe the dominant part of ΔS° to outer-sphere reorganization of the solvent cage induced by the spin change. In particular, they have suggested that "solvent . . . HN hydrogen-bonding interactions, if strongly spin-state dependent, could promote an exceptionally large solvent-sphere reorganization upon spin conversion:"

The dynamics of intersystem crossing processes in solution for 14 six-coordinate d^5, d^6, and d^7 spin-equilibrium metal complexes of iron(III), iron(II), and cobalt(II), for which the spin state lifetimes have been measured by the laser Raman T-jump technique, has extensively been discussed by Sutin, Wilson et al.[190]. The authors have pointed out that the "less spin forbidden" intersystem crossings in the cobalt(II) complexes involving $\Delta S = 1$ changes do not necessarily possess faster crossings than the spin conversion processes in iron(II) and iron(III) complexes, where the spin changes by $\Delta S = 2$. More important, however, should be the potential for term-mixing through spin-orbit coupling. Accordingly, the rate constants for iron(II) and iron-(III) complexes, where no direct spin-orbit coupling occurs between the two spin states involved in the spin state equilibrium (only an indirect mechanism is possible whereby a low-lying spin quartet state in case of iron(III) and a spin triplet in case of iron(II) can mix with both the LS and the HS states separately) are expected to be slower than in cobalt(II) complexes, where spin-orbit coupling between the $^4T(O_h)$ and $^2E(O_h)$ states does occur. There is, however, no obvious correlation between the intersystem crossing rates (and thus the spin state lifetimes) and either metal ion or oxidation state. The fact that "slow" and "fast" processes have been observed within each of the d^5, d^6, and d^7 systems suggests that "nonelectronic" factors can also be significant in determining the intersystem crossing rates. One of these may be the change in the metal-ligand bond distances known to accompany the change in the spin state. As the orbital population fluctuation for cobalt(II) crossover systems $(t_{2g}^6 e_g^* (LS) \rightleftharpoons t_{2g}^5 e_g^{*2} (HS))$ is only half that for iron(II) complexes $(t_{2g}^6 (LS) \rightleftharpoons t_{2g}^4 e_g^{*2} (HS))$ and iron(III) complexes $(t_{2g}^5 (LS) \rightleftharpoons t_{2g}^3 e_g^{*2} (HS))$, the changes in metal-ligand bond lengths ΔR are expected to be smaller in cobalt(II) than in iron(II) and iron(III) crossover systems. Although the enthalpy changes of $\Delta H^{\circ} \approx 2-6$ kcal mol^{-1} observed for spin equilibrium processes are considered to be predominantly due to inner-sphere reorganization, the authors feel that it is not necessarily safe to conclude that the larger geometric changes always imply slower crossing rates. More structural information for spin crossover systems is certainly needed to find out whether there is such a correlation. It is, however, believed[190] that steric hindrance within a ligand system plays an important role in determining intersystem crossing rates. The more flexible the ligand systems, the greater will be the mechanical latitude, the less structural constraint will be imposed on the coordination sphere reorganization, and the faster should be the intersystem crossing rates. This idea receives support from the observed differences in the spin state lifetimes of structurally different spin crossover complexes[190].

10 Brief Survey of Spin Crossover Models

A successful model for the quantitative interpretation of spin transitions should explain all the different types of the temperature dependence of the HS fraction $x_{HS} = \gamma$ as sketched in Fig. 2.

   γ is the fraction of the thermally populated HS states as is reflected in Mössbauer spectra in the slow relaxation limit. The models proposed so far permit to describe one or the other aspect of the transition characteristics. They have the basic assumption in common that the driving force of the spin transition is determined by intramolecular properties of the spin crossover complexes as well as intermolecular interactions communicated through the lattice. Therefore the natural order parameter of the transition is the HS fraction γ which, at a given temperature, is obtained from the minimum of the Gibbs free energy, so that

$$\left(\frac{\partial G}{\partial \gamma}\right)_{p,T} = 0 . \tag{10.1}$$

   To arrive at a common description of the models published so far it is useful to start with the simple expansion of the Gibbs free energy G up to γ^2 terms. In analogy to regular solutions, Slichter and Drickamer[97] have employed the following expression:

$$G = N[(1 - \gamma)G_{LS}(p,T) + \gamma G_{HS}(p,T) + \Gamma(p,T)\gamma(1 - \gamma)] - TS_{mix}. \tag{10.2}$$

$G_{LS}(p,T) = G(p,T,\gamma = 0)$ refers to the Gibbs free energy per molecule of the spin crossover system in the LS state, and correspondingly $G_{HS}(p,T) = G(p,T,\gamma = 1)$ to that of the HS state. N is the number of complex molecules in the system. $\Gamma(p,T)$ is considered as the next term in a power series expansion in the two variables γ and $(1 - \gamma)$. The mixing entropy of both HS and LS molecules present in the lattice writes:

$$S_{mix} = k_B \cdot [N \ln N - N\gamma \ln N\gamma - N(1 - \gamma)\ln N(1 - \gamma)] . \tag{10.3}$$

Equation (10.1) results in an equation for γ:

$$\gamma = e^{-\Delta/kT}/(e^{-\Delta/kT} + e^{-2\Gamma\gamma/kT}) \tag{10.4}$$

where

$$\Delta = (G_{HS} + \Gamma) - G_{LS}.$$

   Slichter and Drickamer have discussed Eq. (10.4) in detail, and they have shown that the general Eq. (10.2) may be appropriate to reproduce the different types of spin conversion curves $\gamma(T)$ of Fig. 2. The various models in the literature differ by the assumptions about G_{LS} and G_{HS}. As $-\Gamma\gamma^2$ can be considered as an interaction term between the HS molecules in the lattice, this equation is also obtained in the molecular mean field approximation used in the theory of magnetism or equivalently in the Ising model solved by the Bragg-Williams approximation. Slichter and Drickamer[97] studied the γ-dependence on (high) pressure and temperature. At ambient pressure the term pdV can be neglected, so that instead of $G(p, T, \gamma)$ the free energy $F(V, T, \gamma) = -kT \ln Z$ with the partition function Z applies.

The model by Chesnut[208], the earliest one which appeared in the literature, simply explains the interaction term Γ by lattice strains. Suppose q_{LS}° and q_{HS}° are equilibrium lattice constants of the pure LS and HS lattice, resp., and q the lattice constant of the real mixture of the spin isomers, then

$$F_\tau(q, T) = F_\tau^\circ(q_\tau^\circ, T) + \frac{1}{2}\xi_\tau(q - q_\tau^\circ)^2, \tau = LS, HS \tag{10.5}$$

represents the first correction for the energy of the lattice. It follows that

$$F = N[F_{LS}(q, T) \cdot (1 - \gamma) + F_{HS}(q, T)\gamma] - T \cdot S_{mix}. \tag{10.6}$$

The mixed lattice has an avaraged ξ value whose dependence on q is neglected. Eq. (10.6) which has no interaction term can be written:

$$F/N = -\frac{1}{2}\xi\, q_{LS}^\circ + \frac{1}{2}\xi[q - (q_{LS}^\circ + \gamma(q_{HS}^\circ - q_{LS}^\circ))]^2 + F_{LS}^\circ(1 - \gamma) + F_{HS}^\circ\gamma +$$

$$+ \frac{1}{2}\xi(q_{LS}^\circ - q_{HS}^\circ)^2\,\gamma(1 - \gamma) - T \cdot S_{mix}/N. \tag{10.7}$$

F reaches its minimum at

$$q_{min} = q_{LS}^\circ + \gamma(q_{HS}^\circ - q_{LS}^\circ) \tag{10.8}$$

which is just the linear interpolation between q_{LS}° and q_{HS}°. $F(q_{min}, \gamma)$ contains the interaction $\Gamma - \frac{1}{2}\xi(q_{LS}^\circ - q_{HS}^\circ)^2$. As was already pointed out by Chesnut the abrupt increase of γ should be accompanied by an abrupt change of the lattice constant q.

The relation between q and γ in Eq. (10.8) allows also to take q as an order parameter. Therefore an equivalent way to obtain γ is calculating first $F(q_{min})$ and then γ taking the Boltzmann distribution. If $E_{LS,HS}^i$ are the energy levels belonging to the HS and LS states of the lattice and if

$$Z = Z_{HS} + Z_{LS} = \sum_i e^{-E_{HS}^i/kT} + \sum_i e^{-E_{LS}^i/kT} \tag{10.9}$$

represents the partition function, then the HS fraction γ becomes

$$\gamma = \frac{Z_{HS}}{Z}. \tag{10.10}$$

Including the lattice strain, the energy changes to

$$E_\tau^i + \frac{1}{2}\xi(q - q_\tau^\circ)^2, \tau = LS, HS \tag{10.11}$$

so that Z is a function of q. $\left(\dfrac{\partial F}{\partial q}\right)_T = 0$ leads to

$$q = (q_{HS}^\circ Z_{HS}(q) + q_{LS}^\circ Z_{LS}(q))/Z(q) \tag{10.12}$$

which, after inserting Eq. (10.10), yields back Eq. (10.8) and is then identical with (10.4).

Chesnut[208] chooses the free energies of the LS and HS lattices as given by the electronic singlet and the g-fold degenerate HS state: $Z_{HS} = g \cdot e^{-E_{HS}/kT}$ and $Z_{LS} = e^{-E_{LS}/kT}$ so that Δ of Eq. (10.4) is $\Delta = (E_{HS} - kT \ln g + \Gamma) - E_{LS}$.

Equation (10.11) is the starting point of the model by Kambara[209,210], who takes the energies $E_{LS,HS}^i$ as the electronic energies of the ligand field Hamiltonian, which, in addition, depends on the strain parameters $q(A_{1g})$, corresponding to the q of Chesnut, and $\{q_u, q_v\}$ transforming as the cubic E_g representation. $E_{LS,HS}^i$ are eigenvalues of the ligand field Hamiltonian

$$\hat{H} = \hat{H}(q_i = 0) + \sum_i \left(\frac{\partial \hat{H}}{\partial q_i}\right)_{q=0} q_i, \qquad i = u, v, A_{1g} \tag{10.13}$$

which includes the Jahn-Teller matrix elements $\left\langle \dfrac{\partial \hat{H}}{\partial q_i} \right\rangle$. This model results in γ-dependent Jahn-Teller distortions.

Zimmermann and König[211] introduce the phonon contribution of the lattice by a Debye model with an interpolated Debye temperature

$$\Theta = \Theta_{LS} + \gamma(\Theta_{HS} - \Theta_{LS}) \tag{10.14}$$

(if $|\Delta\Theta| = |\Theta_{HS} - \Theta_{LS}| \ll 0$) choosen similarly to the behaviour of the q parameter in Eq. (10.8). The free energy $F(\Theta)$ expanded in a power series up to γ^2 gives a temperature dependent interaction term, which changes in a characteristic way the $\gamma(T)$ curves. In addition to the phonon contribution the electronic energy levels of the $^5T_2(O_h)$ state and the singlet $^1A_1$ state are contained in the partition function. The orbital splitting of the $^5T_2$ state is fitted to the temperature dependence of the quadrupole splitting observed by Mössbauer spectroscopy.

To generate sharp transitions the interaction term Γ in Eq. (10.2) must be large. An alternative approach has been used by Sorai and Seki[34] following the theory of heterophase fluctuations. They assume cooperative domains of LS and HS molecules, resp., with a strong interaction within the domains, so that the complexes change spin simultaneously. For n atoms per domain the partition function becomes

$$Z = (Z_{HS}^n + Z_{LS}^n)^{\frac{N}{n}} \tag{10.15}$$

The larger n the steeper is the increase of $\gamma(T)$. Gütlich et al. (Sect. 6.1.1) included in the partition function the vibrational levels of the molecule with an average frequency for the HS state $(\bar{\nu}_{HS})$ and for the LS state $(\bar{\nu}_{LS})$; $\bar{\nu}_{HS} < \bar{\nu}_{LS}$ as observed by FIR spectroscopy.

The models discussed so far have in common that the electronic energies change continuously with the parameter q (see Eq. 10.5) or, which is equivalent, with γ. The change in q can be correlated with a change of the cubic ligand field strength Dq. This correlation has been used in the work of König and Kremer [212,213] who parameterized the experimental susceptibility data by a temperature dependent energy separation $\epsilon(T) = E(^5T_2) - E(^1A_1)$ determined by the ligand field strength.

Wajnflasz's idea [214] deviates significantly from the picture above. He considers a high potential barrier between the HS molecule and the LS molecule, which differ in their radii and their ligand field energy level schemes. The ground states of the two spin isomers are separated by an energy ΔE. In the energy diagram of the LS molecules, the excited HS energy levels are separated by $\epsilon_{LS}(q)$ and, vice versa, the excited LS level is separated by $\epsilon_{HS}(q)$ from the HS ground state in the energy level scheme of the HS molecule. The real fraction of HS molecules becomes different from the fraction γ of the populated HS states. If the exitation energies $\epsilon_{LS,HS}$ are large compared to ΔE and kT, this picture is effectively the same as the above one, where the ϵ parameter then adopts the meaning of ΔE.

There are further models which do not introduce new ideas but use the above frame for more complicated types of interactions. Bari and Sivardière [215] discussed a two sublattice model in analogy to the molecular field theory of antiferromagnetism. In this case there are two different interaction constants, viz. the intrasublattice and the intersublattice interaction. They also expand the one-sublattice model by an Heisenberg type magnetic interaction term between the HS states. Such an interaction may only become important for degenerate spin states.

Ramasesha et al. [216] introduce a coupling between the HS and LS electronic states via lattice strain to take care of the appearance of residual paramagnetism at low temperatures. The resultant mixing of these states is a contradiction to the observations of Mössbauer measurements where the HS and LS states appear separately (slow relaxation limit). Moreover, this mixing cannot occur in the framework of ligand field theory.

Acknowledgement. I wish to express my sincere thanks to my students and research associates, K. Bode, Dr. J. Ensling, Dr. J. Fleisch, Dr. P. Ganguli, Dr. K. M. Hasselbach, H. Köppen, Dr. R. Link, E. W. Müller, I. Sanner, Dr. M. Sorai, Dr. H. Spiering, H. G. Steinhäuser, and G. Sudheimer, who have collaborated with me on spin crossover problems with great enthousiasm. I owe particular thanks to my colleague Dr. H. Spiering, who prepared the chapter on spin crossover models. I also wish to thank Mrs. G. Lehr for typing the manuscript with great care and patience.

We are grateful for the financial help from the *Deutsche Forschungsgemeinschaft*, the *Fonds der Chemischen Industrie*, the *Bundesministerium für Forschung und Technologie*, and the *Alexander von Humboldt Stiftung*, which has enabled us to contribute to the advances in spin crossover research.

11 References

1. Ewald, A. H., et al.: Proc. Roy. Soc. (London) *A 280*, 235 (1964)
2. Ballhausen, C. J.: Introduction to Ligand Field Theory, McGraw-Hill Inc., New York 1962
3. Schläfer, H. L., Gliemann, G.: Einführung in die Ligandenfeldtheorie, Akademische Verlagsges., Frankfurt/M. 1967
4. Cambi, L., Cagnasso, A.: Atti Accad. Naz. Lincei *13*, 809 (1931);
   Cambi, L., Szegö, L.: Ber. dtsch. chem. Ges. *64*, 259 (1931);
   Cambi, L., Szegö, L.: ibid. *66*, 656 (1933);
   Cambi, L., Szegö, L., Cagnasso, A.: Atti Accad. Naz. Lincei *15*, 266 (1932a);
   Cambi, L., Szegö, L., Cagnasso, A.: ibid. *15*, 329 (1932b);
   Cambi, L., Malatesta, L.: Ber. dtsch. chem. Ges. *70*, 2067 (1937)
5. Baker, W. A., Bobonich, H. M.: Inorg. Chem. *3*, 1184 (1964)
6. Lindoy, L. F., Livingstone, S. E.: Coord. Chem. Rev. *2*, 173 (1967)
7. Barefield, E. K., Busch, D. H., Nelson, S. M.: Quart. Rev. *22*, 457 (1968)
8. König. E.: Coord. Chem. Rev. *3*, 471 (1968)
9. Martin, R. L., White, A. H.: in Transiton Metal Chemistry, Vol. 4 (R. L. Carlin (ed.)) Marcel Dekker, Inc., New York 1968, p. 113
10. Sacconi, L.: Pure Appl. Chem. *27*, 161 (1971)
11. König, E.: Ber. Bunsenges. phys. Chem. *76*, 975 (1972)
12. Morassi, R., Bertini, I., Sacconi, L.: Coord. Chem. Rev. *11*, 343 (1973)
13. König, E., Ritter, G.: In: Mössbauer Effect Methodology, Vol. 9 (I. J. Gruverman, C. W. Seidel, D. K. Dieterly (eds.)) Plenum Press, New York–London 1974, p. 3
14. Goodwin, H. A.: Coord. Chem. Rev. *18*, 293 (1976)
15. Gütlich, P.: J. Physique *40* (No. 3) Colloque C2, C2–378 (1979);
    Plenary Lecture, Symposium "Recent Chemical Applications of Mössbauer Spectroscopy", 179th Nat. Meet. Amer. Chem. Soc. Houston (Texas), March 1980
16. Machado, A. A. S. C.: Rev. Port. Quim. *13*, 88 (1971); ibid. *14*, 65 (1972); ibid. *14*, 83 (1972)
17. Drickamer, H. G., Frank, C. W.: Electronic Transitions and the High Pressure Chemistry and Physics of Solids, Chapman and Hall, London 1973
18. Figgis, B. N.: Introduction to Ligand Fields, Interscience Publishers, New York–London–Sydney 1966
19. Earnshaw, A.: Introduction to Magnetochemistry, Academic Press, London–New York 1968
20. Gerloch, M., Slade, R. C.: Ligand-Field Parameters, At the University Press, Cambridge 1973
21. Mabbs, F. E., Machin, D. J.: Magnetism and Transition Metal Chemistry, Chapman and Hall, London 1973
22. van Vleck, J. H.: The Theory of Electric and Magnetic Susceptibilities, Oxford University Press, Oxford 1965
23. Orgel, L. E.: J. Chem. Phys. *23*, 1819 (1955)
24. Griffith, J. S.: J. Inorg. Nucl. Chem. *2*, 1 (1956); ibid. *2*, 229 (1956)
25. Jørgensen, C. K.: Mol. Phys. *2*, 309 (1959); Solid State Phys. *13*, 375 (1962); Mol. Phys. *5*, 271 (1962); Advan. Chem. Phys. *8*, 47 (1965)
26. König, E., Schläfer, H. L.: Z. Physik. Chem (Frankfurt/M.) *34*, 355 (1962)
27. Schäffer, C. E., Jørgensen, C. K.: J. Inorg. Nucl. Chem. *8*, 143 (1958)
28. Jørgensen, C. K.: Absorption Spectra and Chemical Bonding in Complexes, Pergamon Press, Oxford–London–New York–Paris 1962
29. König, E., Kremer, S.: Theoret. Chim. Acta *23*, 12 (1971)
30. Leipoldt, J. G., Coppens, P.: Inorg. Chem. *12*, 2269 (1973)
31. Mikami, M., Konno, M., Saito, Y.: Chem. Phys. Letters *63*, 566 (1979)
32. Ewald, A. H., et al.: Inorg. Chem. *8*, 1837 (1969)
33. Takemoto, J. H., Hutchinson, B.: Inorg. Nucl. Chem. Letters *8*, 769 (1972)

34. Sorai, M., Seki, S.: J. Phys. Chem. Solids *35*, 555 (1974)
35. Hall, G. R., Hendrickson, D. N.: Inorg. Chem. *15*, 607 (1976)
36. Robinson, M. A., Curry, J. D., Busch, D. H.: ibid. *2*, 1178 (1963)
37. Wilson, L. J., Georges, D., Hoselton, M. A.: ibid. *14*, 2968 (1975)
38. König, E., Madeja, K.: ibid. *6*, 48 (1967)
39. Tsipis, C. A., Hadjikostas, C. C., Manoussakis, G. E.: Inorg. Chim. Acta *23*, 163 (1977)
40. Cambi, L., Cagnasso, A.: Atti Accad. Naz. Lincei *19*, 458 (1934)
41. Sugden, S.: J. Chem. Soc. 328 (1943)
42. Burstall, F. H., Nyholm, R. S.: ibid., 3570 (1952)
43. Brandt, W. W., Dwyer, F. P., Gyarfas, E. C.: Chem. Rev. *54*, 959 (1954)
44. Irving, H., Mellor D. H.: J. Chem. Soc. 5222 (1962)
45. Duncan, J. F., Mok, K. F.: ibid. *A*, 1493 (1966)
46. Schilt, A. A.: J. Am. Chem. Soc. *82*, 3000 (1960)
47. Baker, W. A., Jr., Bobonich, H. M.: Inorg. Chem. *2*, 1071 (1963)
48. Madeja, K., König, E.: J. Inorg. Nucl. Chem. *25*, 377 (1963)
49. Collins, R. L., Pettit, R., Baker, W. A., Jr.: ibid. *28*, 1001 (1966)
50. Takemoto, J. H., Streusand, B., Hutchinson, B.: Spectrochim. Acta *30A*, 827 (1974)
51. Beck, W., Schuierer, E.: Chem. Ber. *95*, 3048 (1962)
52. Madeja, K.: Chem. Zvesti *19*, 186 (1965)
53. Basolo, F., Dwyer, F. P.: J. Am. Chem. Soc. *76*, 1454 (1954)
54. Madeja, K., Wilke, W., Schmidt, S.: Z. Anorg. Allgem. Chem. *346*, 306 (1966)
55. Golding, R. M., Mok, K. F., Duncan, J. F.: Inorg. Chem. *5*, 774 (1966)
56. König, E. et al.: Z. Naturforsch. *22A*, 1543 (1967)
57. König, E., Chakravarty, A. S., Madeja, K.: Theoret. Chim. Acta *9*, 171 (1967)
58. Baker, W. A., Jr., Long, G. J.: Chem. Commun. *15*, 368 (1965)
59. König, E., Madeja, K.: Inorg. Chem. *7*, 1848 (1968)
60. König, E., Madeja, K.: J. Am. Chem. Soc. *88*, 4528 (1966)
61. König, E., Kanellakopulos, B.: Chem. Phys. Letters *12*, 485 (1972)
62. König, E., Madeja, K.: Chem. Commun. *3*, 61 (1966)
63. Dézsi, I., et al.: J. Inorg. Nucl. Chem. *29*, 2486 (1967)
64. König, E., Madeja, K.: Inorg. Chem. *6*, 48 (1967)
65. Casey, A. T., Isaac, F.: Aust. J. Chem. *20*, 2765 (1967)
66. König, E., Madeja, K.: Spectrochim. Acta *23A*, 45 (1967)
67. Cunningham, A. J., Fergusson, J. E., Powell, H. K. J.: J. Chem. Soc. (Dalton), 2155 (1972)
68. Ferraro, J. R., Takemoto, J.: Appl. Spectroscopy *28*, 66 (1974)
69. Reiff, W. M., Long, G. J.: Inorg. Chem. *13*, 2150 (1974)
70. Ganguli, P., Gütlich, P., Irler, W., Müller, W.: J. C. S. (Dalton), in press
71. König, E., Madeja, K., Watson, K. J.: J. Am. Chem. Soc. *90*, 1146 (1968)
72. Casey, A. T.: Aust. J. Chem. *21*, 2291 (1968)
73. Driver, R., Walker, W. R.: ibid. *20*, 1375 (1967)
74. Casey, A. T., Thackeray, J. R.: Proc. 12th ICCC, Sydney 1969, p. 9
75. König, E. et al.: J. Phys. Chem. Solids *33*, 327 (1972)
76. König, E., Watson, K. J.: Chem. Phys. Letters *6*, 457 (1970)
77. Takemoto, J. H., Hutchinson, B.: Inorg. Chem. *12*, 705 (1973)
78. Fisher, D. C., Drickamer, H. G.: J. Chem. Phys. *54*, 4825 (1971)
79. König, E., Kremer, S.: Theoret. Chim. Acta *20*, 143 (1971)
80. Greenwood, N. N., Gibb, T. C.: Mössbauer Spectroscopy, Chapman and Hall, London 1971
81. Gütlich, P., Link, R., Trautwein, A.: Mössbauer Spectroscopy and Transition Metal Chemistry, Springer-Verlag, Berlin–Heidelberg–New York 1978
82. König, E., Ritter, G.: Phys. Letters *43A*, 488 (1973)
83. Collins, R. L.: J. Chem. Phys. *42*, 1072 (1965)
84. Sorai, M., Ensling, J., Gütlich, P.: Chem. Phys. *18*, 199 (1976)
85. Gütlich, P., Link, R., Steinhäuser, H. G.: Inorg. Chem. *17*, 2509 (1978)
86. Gütlich, P., Köppen, H., Link, R., Steinhäuser, H.G.: J. Chem. Phys. *70*, 3977 (1979)
87. Sorai, M., Seki, S.: J. phys. Soc. Japan *33*, 575 (1972)

88. König, E. et al.: J. Inorg. Nucl. Chem. *34*, 2877 (1972)
89. Goodwin, H. A., Sylva, R. N.: Aust. J. Chem. *21*, 83 (1968)
90. König, E. et al.: Ber. Bunsenges. phys. Chem. *77*, 390 (1973)
91. Kanellakopulos, B. et al.: J. Physique (Paris) *37* (12), C6–475 (1976)
92. König, E. et al.: J. Phys. C *10*, 603 (1977)
93. Madeja, K. et al.: Z. Anorg. Allg. Chem. *447*, 5 (1978)
94. Greenaway, A. M., Sinn, E.: J. Am. Chem. Soc. *100*, 8080 (1978)
95. König, E., Ritter, G., Kanellakopulos, B.: J. Phys. C *7*, 2681 (1974)
96. König, E., Ritter, G.: Solid State Commun. *18*, 279 (1976)
97. Slichter, C. P., Drickamer, H. G.: J. Chem. Phys. *56*, 2142 (1972)
98. Bargeron, C. B., Drickamer, H. G.: ibid. *55*, 3471 (1971)
99. Spacu, P., Teodorescu, M., Ciomârtan, D.: Mh. Chem. *103*, 1 (1972)
100. Spacu, P. et al.: Z. Anorg. Allg. Chem. *392*, 88 (1972)
101. Irving, H., Mellor, M. J.: J. Chem. Soc. 5237 (1962)
102. Fleisch, J., Gütlich, P., Hasselbach, K. M., Müller, W.: Inorg. Chem. *15*, 958 (1976)
103. König, E. et al.: J. Chem. Phys. *56*, 3139 (1972)
104. Ingalls, R.: Phys. Rev. *133*, 787 (1964)
105. Fleisch, J., Gütlich, P., Hasselbach, K. M.: Inorg. Chem. *16*, 1979 (1977)
106. König, E. et al.: Ber. Bunsenges. physik. Chem. *76*, 393 (1972)
107. König, E., Ritter, G., Goodwin, H. A.: J. Inorg. Nucl. Chem. *39*, 1773 (1977)
108. Gütlich, P., Köppen, H., Sanner, I.: to be published
109. Sylva, R. N., Goodwin, H. A.: Aust. J. Chem. *20*, 479 (1967)
110. Sorai, M., Ensling, J., Hasselbach, K. M., Gütlich, P.: Chem. Phys. *20*, 197 (1977)
111. Halbert, E. J., Harris, C. M., Sinn, E., Sutton, G. J.: Aust. J. Chem. *26*, 951 (1973)
112. Fleisch, J., Gütlich, P., Hasselbach, K. M., Müller, W.: J. Physique (Paris) *35* (12), C6–659 (1974)
113. Fleisch, J., Gütlich, P., Hasselbach, K. M.: Inorg. Chim. Acta *17*, 51 (1976)
114. Fleisch, J.: Dissertation, TH Darmstadt 1976, D17
115. Epstein, L. M.: J. Chem. Phys. *40*, 435 (1964)
116. Collins, R. L., Pettit, R., Baker, W. A., Jr.: J. Inorg. Nucl. Chem. *28*, 1001 (1966)
117. Böhmer, W. H. et al.: ibid. *35*, 1957 (1973)
118. Goodwin, H. A., Sylva, R. N.: Aust. J. Chem. *20*, 217 (1967)
119. Goodwin, H. A., Smith, F. E.: ibid. *25*, 37 (1972)
120. Goodwin, H. A. et al.: ibid. *26*, 521 (1973)
121. Goodwin, H. A., Mather, D. W., Smith, F. E.: ibid. *26*, 2623 (1973)
122. Goodwin, H. A., Smith, F. E.: ibid. *23*, 1545 (1970)
123. Goodwin, H. A., Mather, D. W.: ibid. *27*, 965 (1974)
124. Irler, W. et al.: Solid State Commun. *29*, 39 (1979)
125. Goodwin, H. A., Mather, D. W.: Aust. J. Chem. *27*, 2121 (1974)
126. Goodwin, H. A., Smith, F. E.: Inorg. Nucl. Chem. Lett. *10*, 99 (1974)
127. Goodwin, H. A., Mather, D. W., Smith, F. E.: Aust. J. Chem. *28*, 33 (1975)
128. Mather, D. W., Goodwin, H. A.: ibid. *28*, 505 (1975)
129. Sutton, G. J.: ibid. *13*, 74 (1960)
130. Renovitch, G. A., Baker, W. A., Jr.: J. Am. Chem. Soc. *89*, 6377 (1967)
131. Gütlich, P., Köppen, H., Steinhäuser, H. G.: 74 (3), 475 (1980)
132. Gütlich, P., Sudheimer, G.: to be published
133. Chiswell, B., Lions, F., Mornis, B.: Inorg. Chem. *3*, 110 (1964)
134. Dosser, R. J. et al.: J. Chem. Soc. (A), 810 (1969)
135. Goodgame, D. M. L., Machado, A. A. S. C.: Inorg. Chem. *8*, 2031 (1969)
136. Goodgame, D. M. L., Machado, A. A. S. C.: Chem. Commun. 1420 (1969)
137. Sasaki, Y., Shigematsu, T.: Bull. Chem. Soc. Japan *47*, 109 (1974)
138. Sasaki, Y., Shigematsu, T.: Bull. Chem. Soc. *46*, 3438 (1973)
139. Goodgame, D. M. L.: Bull. Soc. Chim. Fr. 3 (1972)
140. Sams, J. R., Scott, J. C., Tsin, T. B.: Chem. Phys. Letters *18*, 451 (1973)
141. Gosh, S. P., Mishra, L. K.: Inorg. Chim. Acta *7*, 545 (1973)

142. Sams, J. R., Tsin, T. B.: J. Chem. Soc. (Dalton) 488 (1976)
143. Sams, J. R., Tsin, T. B.: Inorg. Chem. 15, 1544 (1976)
144. Goodwin, H. A., Baker, A. T.: Aust. J. Chem. 30, 771 (1977)
145. Reeder, K. A., Dose, E. V., Wilson, L. J.: Inorg. Chem. 17, 1071 (1978)
146. Hennig, H., Benedix, M., Benedix, R.: Z. Chem. 11, 188 (1971)
147. Barth, P., Schmauss, G., Specker, H.: Z. Naturforsch. 27b, 1149 (1972)
148. König, E., Ritter, G., Schnakig, R.: Chem. Phys. Letters 27, 23 (1974)
149. Maeda, Y., Takashima, Y., Nishida, Y.: Bull. Chem. Soc. Japan 49, 2427 (1976)
150. Goodwin, H. A.: Aust. J. Chem. 17, 1366 (1964)
151. Goodwin, H. A., Sylva, R. N.: ibid. 21, 2881 (1968)
152. Goodwin, H. A., Mather, D. W.: ibid. 25, 715 (1972)
153. König, E., Ritter, G., Goodwin, H. A.: Chem. Phys. 1, 17 (1973)
154. Ingalls, R.: Phys. Rev. 133, 787 (1964)
155. Sylva, R. N., Goodwin, H. A.: Aust. J. Chem. 21, 1081 (1968)
156. König, E., Ritter, G., Goodwin, H. A.: Chem. Phys. 5, 211 (1974)
157. Keller, H. J. et al.: Z. Naturforsch. 24b, 1058 (1969)
158. König, E., Ritter, G., Goodwin, H. A.: J. Inorg. Nucl. Chem. 39, 1131 (1977)
159. König, E. et al.: J. Phys. Chem. Solids 39, 521 (1978)
160. König, E., Ritter, G., Goodwin, H. A.: Chem. Phys. Letters 44, 100 (1976)
161. Ritter, G. et al.: Inorg. Chem. 17, 224 (1978)
162. Nelson, S. M., Rodgers, J.: J. Chem. Soc. (A), 272 (1968)
163. Boylan, M. J., Nelson, S. M., Deeney, F. A.: ibid. 976 (1971)
164. Livingstone, S. E., Nolan, J. D.: J. Chem. Soc. (Dalton) 218 (1972)
165. Albert, A., Goldacre, R., Phillip, J.: J. Chem. Soc. 2240 (1948)
166. Bjerrum, J.: Chem. Rev. 46, 381 (1950)
167. Hogg, R., Wilkins, R. G.: J. Chem. Soc. 341 (1962)
168. Kelly, S. J., Ford, G. H., Nelson, S. M.: J. Chem. Soc. (A) 388 (1971)
169. Sacconi, L.: ibid. 248 (1970)
170. Dahlhoff, W. V., Nelson, S. M.: ibid. 2184 (1971)
171. Sacconi, L., Di Vaira, M.: Inorg. Chem. 17, 810 (1978)
172. Jesson, J. P., Weiher, J. F.: J. Chem. Phys. 46, 1995 (1967)
173. Jesson, J. P., Weiher, J. F., Trofimenko, S.: ibid. 48, 2058 (1968)
174. Jesson, J. P., Trofimenko, S., Eaton, D. R.: J. Am. Chem. Soc. 89, 3158 (1967)
175. Beattie, J. K. et al.: ibid. 95, 2052 (1973)
176. Beattie, J. K., Binstead, R. A., West, R. J.: ibid. 100, 3044 (1978)
177. Bradley, G., McKee, V., Nelson, S. M.: J. Chem. Soc. (Dalton) 522 (1978)
178. Robinson, M. A., Busch, D. H.: Inorg. Chem. 2, 1171 (1963)
179. Ganguli, P., Gütlich, P.: to be published
180. Katz, B. A., Strouse, C. E.: J. Am. Chem. Soc. 101, 6214 (1979)
181. Greenaway, A. M. et al.: Inorg. Chem. 18, 2692 (1979)
182. Köppen, H., Link, R., Steinhäuser, H.G., Gütlich, P.: unpublished
183. König, E., Ritter, G., Irler, W.: Chem. Phys. Letters 66, 336 (1979)
184. König, E. et al.: Inorg. Chim. Acta 37, 169 (1979)
185. Ritter, G., König, E., Goodwin, H. A.: J. Inorg. Nucl. Chem. 41, 293 (1979)
186. König, E. et al.: Inorg. Chim. Acta 35, 239 (1979)
187. Levason, W. et al.: J. C. S. Dalton, 1778 (1975)
188. Hoselton, M. A., Wilson, L. J., Drago, R. S.: J. Am. Chem. Soc. 97, 1722 (1975)
189. Hoselton, M. A. et al.: ibid. 98, 6967 (1976)
190. Dose, E. V. et al.: ibid. 100, 1141 (1978)
191. Lazarus, M. S., Hoselton, M. A., Chou, T. S.: Inorg. Chem. 16, 2549 (1977)
192. Goedken, V. L., Merrell, P. H., Busch, D. H.: J. Am. Chem. Soc. 94, 3397 (1972)
193. Dabrowiak, J. C., Merrell, P. H., Busch, D. H.: Inorg. Chem. 11, 1979 (1972)
194. Driessen, W. L., van der Voort, P. H.: Inorg. Chim. Acta 21, 217 (1977)
195. Eibschütz, M., Lines, M. E., DiSalvo, F. J.: Phys. Rev. B15, 103 (1977)
196. Eibschütz, M., DiSalvo, F. J.: Phys. Rev. Lett. 36, 104 (1976)

197. Lines, M. E., Eibschütz, M.: J. Phys. C 9, L 355 (1976)
198. Frank, E., Abeledo, C. R.: Inorg. Chem. 5, 1453 (1966)
199. Richards, R., Johnson, C. E., Hill, H. A. O.: J. Chem. Phys. 48, 5231 (1968)
200. Epstein, L. M., Straub, D. K.: Inorg. Chem. 8, 784 (1969)
201. König, E. et al.: J. Coord. Chem. 2, 257 (1973)
202. Bacci, M. et al.: Inorg. Chem. 12, 1801 (1973)
203. König, E., Ritter, G., Goodwin, H. A.: Chem. Phys. Letters 31, 543 (1975)
204. Gütlich, P., Kläui, W., McGarvey, B. R.: Inorg. Chem., in press
205. El Murr, N., Chaloyard, A., Kläui, W.: ibid. 18, 2629 (1979)
206. Kläui, W.: J. C. S. Chem. Comm. 700 (1979)
207. Rao, P. S., Reuveni, A., McGarvey, B. R., Ganguli, P., Gütlich, P.: Inorg. Chem., in press
208. Chesnut, D. B.: J. Chem. Phys. 40, 405 (1964)
209. Kambara, T.: ibid. 70, 4199 (1979)
210. Kambara, T.: ibid., in press
211. Zimmermann, R., König, E.: J. Phys. Chem. Solids 38, 779 (1977)
212. König, E., Kremer, S.: Theoret. Chim. Acta 20, 143 (1971)
213. König, E., Kremer, S.: ibid. 22, 45 (1971)
214. Wajnflasz, J.: phys. stat. sol. 40, 537 (1970)
215. Bari, R. A., Sivardière, J.: Phys. Rev. B 5, 4466 (1972)
216. Ramasesha, S., Ramakrishnan, T. V., Rao, C. N. R.: J. Phys. C (Solid State Phys.) 12, 1307 (1979)
217. Harris, C. M., Sinn, E., Inorg. Chim. Acta 2, 296 (1968)
218. König, E., Ritter, G., Zimmermann, R.: Chem. Phys. Letters 26, 425 (1974)

12 Appendix

Cross Checking of the Literature on Spin Crossover Research of Iron(II) Complexes

The literature on iron(II) compounds exhibiting spin crossover has been tabulated according to the keywords given below. 'R' in a column means that the corresponding physical method has been applied at room temperature only.

 1 Magnetic susceptibility
 2 Mössbauer Effect
 3 Mössbauer Effect in applied field
 4 Debye-Waller factor
 5 X-ray diffraction (powder)
 6 X-ray diffraction (single crystal)
 7 IR/FIR
 8 UV/vis
 9 NMR
10 ESCA
11 Heat capacity/DTA
12 Relaxation measurements
13 HS/LS in frozen solution
14 HS/LS in liquid solution
15 Effect of ligand replacement
16 Effect of intraligand substitution
17 Effect of noncoordinated anion
18 Effect of crystal solvent
19 Dependence on sample preparation
20 Dependence on time
21 Hydrogen bonding
22 Hysteresis
23 Thermodynamics
24 Theory: LFT description of $\mu(T)$ and other
25 Theory: Spin crossover models
26 Effect of pressure
27 Effect of metal dilution
28 Isotope effect
29 ESR

Compound	1	2	3	4	5	6	7	8	9	10	11	12	13	14	15	16	17	18	19	20	21	22	23	24	25	26	27	28	29	Ref.
[Fe(phen)₂(NCS)₂]	X				X									X	X															5
		R				R	R	R						X	X															58
	R	X			X		X	X						X	X									X						45
	X	X		X			X							X	X									X						62
		X			X				X					X																55
	X	X					X							X											X					49
		X					X	X											X	X										65
												X		X	X	X	X	X	X											56
	X	X												X	X									X		X				66
		X	X																											63
		X												X	X			X	X					X						64
	X	X					X	X		X	X							X												78
	X	X					X	X						X	X	X							X							33
		X			X		X	X						X	X	X									X					67
	X	X		X		R	X							X	X			X	X			X		X						34
		X												X	X							X								68
	X	X			X		X				X			X	X				X											50
		X												X	X															69
	X																													70
																					X						X			179
[Fe(R-phen)₂(NCS)₂]																														
R = 5–Cl, 5–NO₂, 5–CH₃, 5–C₆H₅	X	X															X	X	X											67
mono- and polysubstituted	X	X	X											X											X					98
R = 4,7–(CH₃)₂	X	X	X	X																				X						218
	X	X		X																	X									95
	X																													96
R = 4–CH₃	X	X		X										X	X									X						211
	X	X	X	X										X	X															91
	X	X																					X							92
various 4- and 4,7-substituted	X	X																												90
	X																													93

Compound	1	2	3	4	5	6	7	8	9	10	11	12	13	14	15	16	17	18	19	20	21	22	23	24	25	26	27	28	29	Ref.
[Fe(phen)₂(NCSe)₂]	X	X			X										X															5
	X	X			X																									62
							X	X							X											X				56
		X			X		X	X							X									X						66
							X	X			X														X					64
							X	X							X								X		X					78
					X		X	X																	X					34
																										X				68
															X															77
[Fe(bipy)₂(NCS)₂]	X	X																												5
		X					X								X															58
		X					X								X											X				56
							X	X							X															68
	X	X				R	R																							50
	X	X					X	X																						49
				X		X													X	X										71
	X														X				X	X									X	72
	X																													76
							X																							82
																														77
[Fe(4,4'-(CH₃)₂-bipy)₂(NCSe)₂]	X														X															67
[Fe(4,4'-(CH₃)₂-bipy)₂(NCS)₂]	X	X													X															67
[Fe(py)₂(phen)(NCS)₂]	X	X																												100
																														99
[Fe(pip)₂(NCS)₂]	X	X																												93

Compound	1	2	3	4	5	6	7	8	9	10	11	12	13	14	15	16	17	18	19	20	21	22	23	24	25	26	27	28	29	Ref.
[Fe(2-Y-phen)₃] X₂																														
Y = CH₃, OCH₃ X = ClO₄⁻	X	X					X								X															112
Y = CH₃ X = ClO₄⁻		X																												102
Y = CH₃ X = BF₄⁻	X	X										X																		107
Y = CH₃ X = BF₄⁻, PF₆⁻, ClO₄⁻, I⁻	X	X						R					X			X							X			X				89
Y = CH₃ X = ClO₄⁻, I⁻		X														X														98
Y = CH₃ X = ClO₄⁻, BF₄⁻	X	X	X																					X						103
Y = CH₃ X = BPh₄⁻, I⁻	X	X	X				X										X							X						106
Y = OCH₃ X = ClO₄⁻	X	X													X															105
[Fe(phen)₂X] · 5 H₂O																														
X = C₂O₄²⁻, CH₂C₂O₄²⁻	X	X	X				R							X			X													186
[Fe(2-R-phen)₂] X₂																														
R = CH=N–R', X = BF₄⁻	X	X						R							X	X														125
R = (structure, N–CH₃ oxazole), X = BF₄⁻, ClO₄⁻, I⁻	X															X				X										126
R = (thiazole structure), X = BF₄⁻	X														X															127
R = (thiazolidine structure), X = BF₄⁻	X														X															127

Compound	1	2	3	4	5	6	7	8	9	10	11	12	13	14	15	16	17	18	19	20	21	22	23	24	25	26	27	28	29	Ref.
(benzothiazole structure) R = ... X = BF_4^-	X															X														127
(imidazoline structure) R = ... X = $BF_4^-(H_2O)$	X							R																						128
(benzimidazole structure) R = ... X = $BF_4^-(2\,H_2O)$	X							R																						128
(benzimidazole anion structure) R = ... X = ClO_4^-	X							R																						128
R = CH=N–NR′R″ X = ClO_4^-	X				X	R										X														123 124
[Fe(2-pic)$_3$] X$_2$ · Sol X = Cl$^-$, Br$^-$, I$^-$	X	X			R												X			X				X		X	X			130
X = Cl$^-$ Sol = EtOH		X																				X		X	X					35
X = Cl$^-$ Sol = EtOH, MeOH, H_2O, 2 H_2O		X																			X		X		X	X				39
X = Cl$^-$ Sol = EtOH		X																			X		X		X		X			85
X = Cl$^-$ Sol = EtOH																														86
X = Cl$^-$ Sol = MeOH, 2 H_2O	X					R											X			X										94
X = Cl$^-$ Sol = EtOH						X														X			X							31
X = Cl$^-$ Sol = EtOH, MeOH	X					X											X	X	X	X			X							180
X = Br$^-$ Sol = EtOH, MeOH	X																X	X	X	X										181
X = Cl$^-$ Sol = EtOH	X																X		X	X										181
X = Cl$^-$ Sol = C_2H_5OD, CH_3OD		X				R																								131, 132

Compound	1	2	3	4	5	6	7	8	9	10	11	12	13	14	15	16	17	18	19	20	21	22	23	24	25	26	27	28	29	Ref.
[Fe(2-R-py)$_3$] X$_2$																														
R = (pyim) X = ClO$_4^-$(H$_2$O)	R																													133
various X	X	X						R									X							X						134
X = ClO$_4^-$(H$_2$O)	X	X						R								X							X	X						135
X = ClO$_4^-$(H$_2$O)	X	X						X					X	X										X						138
X = BPh$_4^-$	X							X							X		X	X						X						190, 145
R = (pyimi) X = ClO$_4^-$	X	X																X	X											136
R = (pyben) various X	X	X					R	R								X	X			X	X									140
X = Br$^-$, NO$_3^-$(H$_2$O) = ClO$_4^-$ (2 H$_2$O)	X	X	X				R	R							X	X	X			X	X									143
various X	X	X	X					X							X	X							X	X						142
X = BF$_4^-$(3 H$_2$O), BF$_4^-$	X	X						R								X							X	X						144
X = ClO$_4$(H$_2$O)	X	X						R																						138
R = (Ph-pyrazole) X = ClO$_4^-$	X																													146

Compound	1	2	3	4	5	6	7	8	9	10	11	12	13	14	15	16	17	18	19	20	21	22	23	24	25	26	27	28	29	Ref.
$[Fe(2\text{-}R\text{-}py)_2(NCS)_2]$ R = CH=N–R' various R'		X					R	R								X							X							149
	X	X					R	R								X														147
R' = Ph	X							R																						148
$[Fe(pythiaz)_2]\,X_2$	X	X		X				R															X							151
pythiaz =	X	X	X	X																			X	X						11
X = ClO_4^-, BF_4^-		X	X																					X						153
$[Fe(paptH)_2]\,X_2$	X	X						R				X					X	X		X		X	X							109
paptH =		X	X														X	X				X	X							217
various X	X	X											X				X	X		X			X	X						160
X = Cl^- (2 H_2O)		X												X				X						X						176
X = NO_3^-																					X	X	X	X						161
$[Fe(papt)_2]$	X	X		X			X											X												155
	X	X		X														X												158
	X	X																X												159
	X		X				X		X																					157
$[Fe(papt)_2]\cdot Sol$ Sol = C_6H_6, 4/3 $CHCl_3$	X	X	X	X														X	X			X								156

Compound	1	2	3	4	5	6	7	8	9	10	11	12	13	14	15	16	17	18	19	20	21	22	23	24	25	26	27	28	29	Ref.
[Fe(R-papt)₂]·Sol R = 3-CH₃ Sol = 1/6 CHCl₃ R = 4-CH₃ Sol = CHCl₃		X	X									X											X							185
[FeL₂](ClO₄)₂·3 H₂O L =	X							R																						164
[Fe(ppa)₂]X₂ n·H₂O various X, H₂O ppa =	X	X						R							X	X					X	X	X							163
[Fe(pnp) XY] pnp = X = Y = I; X = I, Y = NCS	X							R						X	X								X							168
[FeLCl₂]·H₂O	X	X	X				R	R																						201
L =	X	X				R	R							X																120

Compound	1	2	3	4	5	6	7	8	9	10	11	12	13	14	15	16	17	18	19	20	21	22	23	24	25	26	27	28	29	Ref.
[FeX(pppp)] BPh$_4$ pppp = Ph$_2$P–CH$_2$–CH$_2$–PPh–CH$_2$CH$_2$–PPh–CH$_2$–PPh$_2$																														
X = Cl, Br, I	X																													171
X = Br, I		X													X															203
[Fe(HB(X, Y-pz)$_3$)$_2$] X, Y–pz =																														
X = CH$_3$, Y = H	X	X						X	X				X		X	X								X						172
X = Y = H	X	X							X					X		X							X	X						174
X = CH$_3$, Y = H	X	X									X		X	X							X	X	X	X						173
X = Y = H	X												X	X							X	X								175
							R	R																						190, 176
[Fe(btz)$_2$(NCS)$_2$] btz =	X																													177

Compound	1	2	3	4	5	6	7	8	9	10	11	12	13	14	15	16	17	18	19	20	21	22	23	24	25	26	27	28	29	Ref.
[Fe(R₂bt)₂X₂]																														
R = H, X = NCS, NCSe	X															X						X								177
R = CH₃, X = NCS		X			X														X			X	X		X					184
R = H, X = NCS					X																									124
[Fe(dippen)₂Cl₂] dippen = Ph₂P–CH=CH–PPh₂	X	X													X															187
[FeL](PF₆)₂																														
II: R = R' = H, R'' = CH₃	X	X					R	R	X				X		X															37
III: R = H, R' = R'' = CH₃	X	X				X	R				X	X		X	X								X	X						188
IV: R = R' = R'' = CH₃										X																				189
II, III, IV												X				X														191
II, III												X			X	X								X	X					190

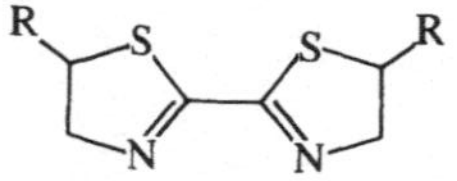

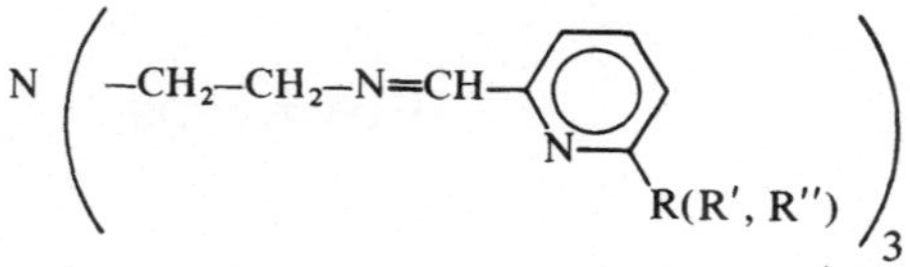

R₂bt =

[FeL](PF₆)₂

L = N(—CH₂—CH₂—N=CH—pyridyl(R, R', R''))₃

Compound	1	2	3	4	5	6	7	8	9	10	11	12	13	14	15	16	17	18	19	20	21	22	23	24	25	26	27	28	29	Ref.
[Fe([14] diene N$_4$) (phen)$_2$] (ClO$_4$)$_2$	X						R								X								X							192
[Fe([14] ane N$_4$) (NCS)$_2$]	X						R	R	X						X								X							193
[Fe(isoxazole)$_6$] X$_2$ X = BF$_4^-$, ClO$_4^-$	X	X			R		R																							194
1 T−Fe$_x$Ta$_{1-x}$S$_2$ (X $\leqslant$ 1/3)	X	X																							X					196
	X	X																												195

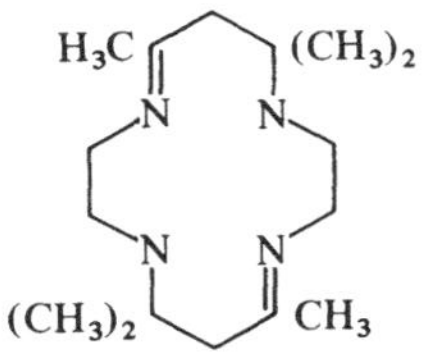

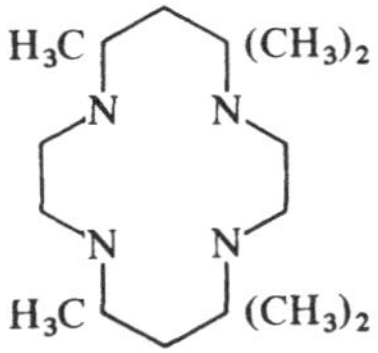

[14] diene N$_4$ =

[14] ane N$_4$ =

Structure and Bonding

Editors: J. D. Dunitz, J. B. Goodenough,
P. Hemmerich, J. A. Ibers, C. K. Jørgensen,
J. B. Neilands, D. Reinen, R. J. P. Williams

Volume 40
Biochemistry

1980. 35 figures, 14 tables. VII, 146 pages
ISBN 3-540-09816-X

Contents:
I. A. Cohen: Metal-Metal Interactions in Metalloporphyrins, Metalloproteins and Metalloenzymes (153 ref.)
L. Que, Jr.: Non-Heme Iron Dioxygenases. Structure and Mechanism (105 ref.)
H. Umezawa, T. Takita: The Bleomycins: Antitumor Copper-Binding Antibiotics (57 ref.)
W. Rüdiger: Phytochrome, A Light Receptor of Plant Photomorphogenesis (198 ref.)

Volume 41
Molecular Structure and Sensory Physiology

1980. 57 figures, 3 tables. VII, 146 pages
ISBN 3-540-09958-1

Contents:
W. Schmidt: Physiological Bluelight Reception (203 ref.)
M. J. Doughty, B. Diehn: Flavins as Photoreceptor Pigments for Behavioral Responses in Motile Microorganisms, Especially in the Flagellated Alga, Euglena sp (111 ref.)
V. E. A. Russo, P. Galland: Sensory Physiology of Phycomyces blakesleeanus (94 ref.)
W. Nultsch, D.-P. Häder: Light Perception and Sensory Transduction in Photosynthetic Prokaryotes (139 ref.)

Volume 42
Luminescence and Energy Transfer

1980. 64 figures, 10 tables. V, 133 pages
ISBN 3-540-10395-3

Contents:
G. Blasse: The Luminescence of Closed-Shell Transition-Metal Complexes. New Developments. – R. C. Powell, G. Blasse: Energy Transfer in Concentrated Systems. – K. C. Bleijenberg: Luminescence Properties of Uranate Centres in Solids.

Volume 43
Bonding Problems

1980. Approx. 58 figures, approx. 30 tabels.
Approx. 240 pages
ISBN 3-540-10407-0

Contents:
C. K. Jørgensen: The Conditions for Total Symmetry Stabilizing Molecules, Atoms, Nuclei and Hadrons. – J. C. Green: Gas Phase Photoelectron Spectra of d- and f-Block Organometallic Compounds. – R. Englman: Vibrations in Interaction with Inpurities. – W. L. Smith, K. N. Raymond: Actinide-Specific Sequestering Agents and Decontamination Applications. – Y. Y. G. Moura, I. Moura, A. V. Xavier: Novel Structures in Iron-Sulfur Proteins.

B. Rånby, J. F. Rabek
ESR Spectroscopy in Polymer Research

1977. 356 figures, 29 tables. XIV, 410 pages
(Polymers/Properties and Applications, Volume 1)
ISBN 3-540-08151-8

Contents:
Generation of Free Radicals. – Principles of ESR Spectroscopy. – Experimental Instrumentation of Electron Spin Resonance. – ESR Study of Polymerization Processes. – ESR Study of Degradation Processes in Polymers. – ESR Study of Polymers in Reactive Gases. – ESR Studies of the Oxidation of Polymers. – ESR Studies of Molecular Fracture in Polymers. – ESR Studies of Graft Copolymerization. – ESR Studies of Crosslinking. – Application of Stable Free Radicals in Polymer Research. – ESR Spectroscopy of Stable Polymer Radicals and their Low Molecular Analogues. – ESR Study of Ion-Exchange Resins.

Springer-Verlag
Berlin
Heidelberg
New York

Crystals

Growth, Properties, and Applications

Managing Editor: H. C. Freyhardt

Volume 1
Crystals for Magnetic Applications

Editor: C. J. M. Rooijmans
1978. 79 figures, 8 tables. VI, 139 pages
ISBN 3-540-09002-9

Contents:

Volume 2
Growth and Properties

Editor: H. C. Freyhardt
1980. 97 figures, 23 tables. VI, 199 pages
ISBN 3-540-09697-3

Contents:

Volume 3
III-V Semiconductors

1980. 92 figures, 9 tables. VI, 163 pages
ISBN 3-540-09957-3

Contents:

Volume 4
Organic Crystals. Germanates. Semiconductors

1980. 133 figures, 23 tables. V, 221 pages
ISBN 3-540-10298-1

Contents:

Springer-Verlag
Berlin
Heidelberg
New York